ON EARTH AS IN HEAVEN

Orthodox Christianity and Contemporary Thought
Series Editors: George Demacopoulos and Aristotle Papanikolaou

This series consists of books that seek to bring Orthodox Christianity into an engagement with contemporary forms of thought. Its goal is to promote (1) historical studies in Orthodox Christianity that are interdisciplinary, employ a variety of methods, and speak to contemporary issues; and (2) constructive theological arguments in conversation with patristic sources and that focus on contemporary questions ranging from the traditional theological and philosophical themes of God and human identity to cultural, political, economic, and ethical concerns. The books in the series will explore both the relevancy of Orthodox Christianity to contemporary challenges and the impact of contemporary modes of thought on Orthodox self-understandings.

ON EARTH
AS IN HEAVEN

Ecological Vision and Initiatives of
Ecumenical Patriarch Bartholomew

ECUMENICAL PATRIARCH BARTHOLOMEW

Edited and with an Introduction by
JOHN CHRYSSAVGIS

FORDHAM UNIVERSITY PRESS
New York 2012

Library of Congress Cataloging-in-Publication Data

Bartholomew I, Ecumenical Patriarch of Constantinople, 1940–
  On earth as in heaven : ecological vision and initiatives of Ecumenical Patriarch Bartholomew / Ecumenical Patriarch Bartholomew ; edited and with an introduction by John Chryssavgis.—1st ed.
    p. cm.— (Orthodox Christianity and contemporary thought)
    Includes index.
    ISBN 978-0-8232-3885-9 (cloth : alk. paper)
    1. Human ecology—Religious aspects—Orthodox Eastern Church.   2. Environmental degradation—Religious aspects—Christianity.   3. Global warming—Religious aspects—Christianity.   4. Climatic changes.   5. Orthodox Eastern Church—Doctrines.   I. Chryssavgis, John.   II. Title.
BX337.5.B368   2012
261.8′80882819—dc23

                                                                                2011018517

14 13   5 4 3 2
First edition

Contents

Foreword

There is a law of unintended consequences, for when things go wrong; but there should also be a law for unexpected consequences when things happen to go right. While I was International President of the World Wide Fund for Nature in the 1980s, one of our main concerns was how to get it across to the public that there really was a serious threat to the survival of many wild species of plants and animals and to the wild areas that are their habitats. It seemed to me that while books, lectures, newspaper articles, and television programs were essential to getting the message across, they reached only those members of the public who used these means of getting to know what was going on in the world around them. These are members of the educated and literate group whose "world around them" is either a city or an advanced agricultural economy. On the other other hand, the inhabitants of the great wilderness areas of the world are less likely to be literate and very unlikely to have the benefit of printed and electronic media.

It seemed to me that the only people who might be in a position to know about the threats to the natural environment and to be in touch with the people who lived and survived in the areas where wildlife was most as risk would be their local religious leaders. It then struck me that most, if not all, religious faiths acknowledged that our universe was the creation of an Almighty being—in which case, each of these faiths ought to feel a responsibility to care for the "creation" of our planet Earth with its great expanses of wilderness lands and for the great multitude of wild animals and plants.

It so happened that 1986 was the twenty-fifth anniversary of the founding of WWF, and there was much discussion about where to hold the Annual General Meeting and the celebrations of that anniversary. After much discussion, it was eventually decided to hold the meeting at Assisi, the home of St. Francis, whose concern for wild animals is well known. It was also agreed that it should be suggested to the Minister General of the Franciscan Order that he might invite leaders of the other major faiths to a meeting in Assisi to discuss among themselves their several attitudes toward God's creation and the natural world. There was no suggestion that they should agree amongst themselves, only that they should each attempt to define their particular traditional beliefs about "the creation."

At the end of the meeting, it was decided to hold a press conference to try to explain what the faith leaders had been discussing. When one journalist asked whether it was intended to set up a new organization, I found myself replying to the effect that I did not think that any formal organization was necessary, but that WWF would happily act as a "consultant" to the faiths on any practical questions of conservation that they might care to raise. Out of this grew the concept of the Alliance of Religions and Conservation—which, conveniently, became ARC for short. This was the unexpected consequence of my speculation.

In the event, Assisi posed a challenge to the major faiths, and each responded in its own way, but none with greater enthusiasm than the Greek Orthodox Church under the inspired leadership of His All Holiness the Ecumenical Patriarch of Constantinople. In September 1988, I was invited to an interfaith gathering initiated by the Ecumenical Patriarchate on the island of Patmos, where St. John wrote his Revelation, to discuss the subject of "Religion and the Environment." Unable to attend in person, I later proposed that the then-Ecumenical Patriarch consider establishing a special day of reflection and celebration of the world's natural heritage. This was agreed to by the Holy Synod, and the first such day was commemorated on September 1, 1989. That meeing proved to be only the first of a long series of activities and meetings with respect to the environment by the Ecumenical Patriarchate.

In 1991, His All Holiness Bartholomew was elected Patriarch of Constantinople, and, through his efforts, September 1 has since become the focus for Orthodox, and now Protestant Christian, reflection and action to protect the world's natural environment. The writings of His All Holiness on this subject for successive celebrations of September 1 open this significant volume. His work has not just involved a remarkable range of lectures, presentations and speeches—as can be seen in this volume—he has also sought to bring people of different faiths together to think about how best to work together to halt the rapid degradation of the world's natural environment.

WWF also helped to fund the first series of theological meetings held on the island of Halki in 1995, which I was able to attend. Since then, His All Holiness has been actively involved in the development of conservation programs by Orthodox Christians all over the world.

It is only too evident that this commitment is due to his exploration and understanding of the insights and teachings of his Church. He is not following a fashion; His All Holiness quite evidently cares for nature because Holy Scripture tells him to do so, because the Church fathers show him this is right, and because Orthodox tradition has retained and kept alive the belief that we are part of the greater purpose of God for all life on Earth.

For more than twenty years, Ecumenical Patriarch Bartholomew has spoken about the Christian relationship with nature, and, as many of the speeches in this book testify, he has taken the time to delve deeply into the philosophical, psychological, and theological issues that are raised by the question of our place within God's purpose, the purpose of nature, and to reflect upon what it means to be human within this broader context.

In 1986 I had a vague idea about trying to encourage the world's faiths to take an interest in the protection of the world's natural environment, and, possibly, to become allies with the world's conservation movements. The dedicated commitment of His All Holiness, as reflected in this book, has been the unexpected consequence of that tentative conception, and I have nothing but deep admiration and gratitude for what His All Holiness has achieved.

Philip
Duke of Edinburgh
Buckingham Palace

PREFACE

This is the third and final volume of Ecumenical Patriarch Bartholomew's selected writings. The first volume spoke to a contemporary world about human rights, religious tolerance, and international peace. The second volume contained major addresses relating to the ecumenical vision of His All Holiness and his sincere desire "for the unity of the Churches of God" through dialogue with other Christian confessions. The entire series accompanies his earlier monograph entitled *Encountering the Mystery: Understanding Orthodox Christianity Today* (New York: Doubleday, 2008).

On Earth as in Heaven contains the Ecumenical Patriarch's statements on environmental degradation, global warming, and climate change. It includes numerous speeches and interviews in various circumstances, such as ecological symposia and academic seminars in regional and international events over the first twenty years of his ministry.[1] Finally, this volume also encompasses a selection of pastoral letters and exhortations—ecclesiastical, ecumenical, and academic in nature—by His All Holiness for a variety of occasions, such as Easter and Christmas, honorary doctorates, and academic awards.[2]

Once again, I am grateful for the tireless guidance received from the staff at Fordham University Press, which has produced a striking and substantial series, as well as for the ongoing support of Dr. Aristotle

---

1. As in earlier volumes, subtitles and footnotes throughout the introduction and the volume belong to the editor and are provided to facilitate reference and reading of the texts.

2. Some of the texts in the present volume originally appeared in *Cosmic Grace, Humble Prayer: Ecological Vision of the Green Patriarch Bartholomew* (Grand Rapids, Mich.: Eerdmans Publications, 2003 and [revised] 2009), which is no longer in print. Portions of the introduction also appeared in the *International Journal of Environmental Studies* 64, 2007, 9–18. Reprinted with the kind permission of Taylor and Francis.

Papanikolaou and Dr. George Demacopoulos, co–founding directors of the Orthodox Christian Studies Program at Fordham University.

I would be remiss if I did not publicly acknowledge that the concept of this series of books belongs to the Very Rev. Alexander Karloutsos, to whom I remain personally indebted for the opportunity and blessing to contribute to its realization.

Finally—although foremost of all—I have been profoundly humbled and honored that His All Holiness has so graciously and generously as-signed and entrusted to me the task of selecting, editing, and introducing his written texts through this series of three volumes, which have proved for me an endless source of inspiration while also providing readers with an insight into the magnitude and magnificence of his vision and leadership.

J.C.

ON EARTH AS IN HEAVEN

Introduction

REV. DR. JOHN CHRYSSAVGIS

In the past two decades, the world has witnessed alarming environmental degradation—with climate change, the loss of biodiversity, and the pollution of natural resources—and the widening gap between rich and poor, as well as increasing failure to implement environmental policies. During the same decade, one religious leader has discerned the signs of the times and called people's attention to this ecological and social situation. The worldwide leader of the Orthodox Churches, His All Holiness Ecumenical Patriarch Bartholomew has persistently proclaimed the primacy of spiritual values in determining environmental ethics and action.

No worldwide church leader has been as recognized internationally for his dynamic leadership and initiatives in addressing the theological, ethical, and practical imperative in relation to the critical environmental issues of our time as has His All Holiness Ecumenical Patriarch Bartholomew. Patriarch Bartholomew has long placed the environment at the head of his church's agenda by developing ecological programs, chairing Pan-Orthodox gatherings and international symposia, and organizing environmental seminars for more than a decade.

THE GREEN PATRIARCH

Patriarch Bartholomew (born Demetrios Archontonis), the current Ecumenical Patriarch of the Orthodox Church, was born on February 29, 1940, in a small village on the island of Imvros, modern-day Gökçeada in Turkey. The residents of this island, like the inhabitants of so many other regions of Asia Minor, have been known through the centuries for their profound tradition and pious devotion, and in general for their cultivation of spiritual values. As the young Demetrios, Patriarch Bartholomew was raised to work the earth of the heart, long before he would be prepared to preserve the green of the environment as Ecumenical Patriarch.

His theological training also attracted the young Demetrios to move beyond the library and to breathe the air of the *oikoumene*, the breadth of the universe of theological communication and ecclesiastical reconciliation. In later years, he would see a similar connection between church and environment:

> For us at the Ecumenical Patriarchate, the term "ecumenical" is more than a name: it is a worldview, and a way of life. The Lord intervenes and fills His creation with His divine presence in a continuous bond. Let us work together so that we may renew the harmony between heaven and earth, so that we may transform every detail and every element of life. Let us love one another. With love, let us share with others everything we know and especially that which is useful in order to educate godly persons so that they may sanctify God's creation for the glory of His holy name.[1]

In 1961, Demetrios graduated from the Patriarchal School at Halki,[2] which for 127 years trained numerous clergymen and theologians of the Ecumenical Patriarchate throughout the world until it was forced to close its doors officially in 1971. This school was to be—and still remains—the

---

1. See his address at Scenic Hudson (November 13, 2000).

2. The Patriarchate's international theological school at Halki (Heybeliada, on the Princes' Islands in the Sea of Marmara) has been closed since 1971, further to a Turkish law forbidding private universities to function. The closure would appear to be in breach of Article 40 of the Treaty of Lausanne and Article 9 of the European Convention on Human Rights. Yet Halki served as the formative and theological center for numerous leaders of the (especially, but not only) Greek-speaking Orthodox world. The function of Halki had been diminished both as a secondary school and graduate seminary since the late 1950s. The magnificent 19th-century building contains a library of 40,000 books and historical manuscripts, as well as classrooms filled with old wooden desks, and spacious reception and dormitory rooms. It is Patriarch Bartholomew's dream and desire to reopen the Theological School, where he often retires to rest and write.

venue for numerous meetings and seminars on environmental issues during his tenure as Patriarch. In 1961, Demetrios was ordained to deacon. It was during this office that the young Demetrios also received the monastic name of Bartholomew.

From 1963 to 1968, Bartholomew attended several prestigious centers of scholarship and ecumenical dialogue, such as the University of Munich, the Ecumenical Institute in Bossey (Switzerland), and the Institute of Oriental Studies of the Gregorian University in Rome. In the last of these institutions, he received his doctorate in Canon Law. His numerous publications covered such diverse subjects as the role of the Ecumenical Patriarchate, the apostolic mission of the Orthodox Church, the tradition and witness of the Church in the contemporary world, theological dialogues with the Roman Catholic Church, and ecumenical relations with other Christian confessions.

Upon the completion of his studies, Bartholomew returned to Constantinople, where, in 1969, he was ordained to the priesthood. From 1972, he served as director of the newly established special Personal Office of the late Patriarch Dimitrios, and in 1973 he was elected Metropolitan of Philadelphia. He attended General Assemblies of the World Council of Churches—even serving as Vice-Chairman of the Faith and Order Commission as well as a member of its Central and Executive committees—from 1968 to 1991. These were critical and formative years for the development of the ecological sensitivity of this influential international organization. In 1990, Bartholomew was elected Metropolitan of Chalcedon, the most senior rank among the bishops in Constantinople at the time. And in October 1991, he was elected Ecumenical Patriarch, the most senior of Orthodox bishops throughout the world, "first among equals" among all Orthodox Patriarchs and Primates.

From the outset of his tenure—indeed, from the very moment of his enthronement address—Patriarch Bartholomew outlined the dimensions of his leadership and vision within the Orthodox Church: the vigilant education in matters of theology, liturgy, and spirituality; the strengthening of Orthodox unity and cooperation; the continuation of ecumenical engagements with other Christian churches and confessions; the intensification of inter-religious dialogue for peaceful coexistence; and the initiation of discussion and action for the protection of the environment against ecological pollution and destruction.

Perhaps no other church leader in history has emphasized ecumenical dialogue and communication as a primary intention of his tenure. Certainly, no other church leader in history has brought environmental issues to the foreground, indeed to the very center of personal and ecclesiastical attention. His initiatives and endeavors have earned Patriarch Bartholomew the title "Green Patriarch"—a title formalized in the White House in 1997 by Al Gore, Vice President of the United States—and several significant environmental prizes. In 2000, Scenic Hudson honored the Ecumenical Patriarch with the International Visionary Award for Environmental Achievement. *The Guardian* in London recognized him as one of the world's leaders in raising environmental awareness. In 2008, Ecumenical Patriarch Bartholomew was named one of *Time* magazine's 100 Most Influential People in the World for "defining environmentalism as a spiritual responsibility."

INITIATIVES AND ACTIVITIES

The environmental vision and initiatives of the Ecumenical Patriarchate date to the mid-1980s, when it organized and chaired the third session of the Pre-Synodal Pan-Orthodox Conference in Chambésy (October 28– November 6, 1986). Although the decisions of this meeting were not binding, serving only as recommendation, nevertheless the representatives attending the meeting expressed and stressed their concern for the abuse of the natural environment by human beings, especially in affluent societies of the Western world. The meeting also underlined the importance of respecting the sacredness and freedom of the human person created in the image and likeness of God, the missionary imperative and witness of the Orthodox Church in the contemporary world, as well as the harm wrought by war, racism, and inequality on human societies and the environment. The emphasis was on leaving a better world for future generations.

Thereafter, and especially as a result of the General Assembly of the World Council of Churches held in Vancouver (1983), several Inter-Orthodox meetings were organized on the subject of "Justice, Peace, and the Integrity of Creation" and attended by Orthodox representatives. Three significant consultations were held: the first was in Sofia, Bulgaria (1987).

A second consultation, in Patmos, Greece (1988), marked the 900th anniversary of the foundation of the historic Monastery of St. John the Theologian on the island of Patmos. This attractive and historic island in the Mediterranean will later become the focus of a new initiative of the current Ecumenical Patriarch. However, on September 23–25, 1988, Patriarch Dimitrios, under whose spiritual jurisdiction the Monastery lies, assigned Metropolitan John (Zizioulas) of Pergamon as the Patriarchal representative to this conference on the occasion of this anniversary and organized by the Greek Ministry of Cultural Affairs in cooperation with the local civil authorities.[3] One of the primary recommendations of this conference was that the Ecumenical Patriarchate should assume the responsibility of appointing a particular day of the year as especially designated and dedicated for the protection of the natural environment.

This conference proved to be a catalyst for the direction of many subsequent Patriarchal initiatives on the environment. The Christmas encyclical letter, signed by Patriarch Dimitrios, already looked forward to the Patmos celebrations and symposium of 1995. A third Inter-Orthodox consultation was held in Minsk, Russia (1989), while an environmental program was also piloted in Ormylia (1990).

In 1989, the same Ecumenical Patriarch Dimitrios, immediate predecessor of Patriarch Bartholomew, who was his closest theological and administrative advisor in these and other matters, published the first official decree, an encyclical letter sent out "to the *pleroma* of the Church."[4] Dimitrios was known for his softness and meekness. Therefore, it always seemed so fitting that it was during his tenure that the Orthodox Church worldwide was invited to dedicate a day of prayer for the protection of the environment, which human beings have treated so harshly; Dimitrios encouraged his faithful to walk gently on the earth, just as he had. This encyclical, proclaimed on the occasion of the first day of the new ecclesiastical calendar, known as the *indictus*, formally established September 1st

---

3. The Patmos conference was entitled "Religions, the Material World and the Natural Environment." Unfortunately, neither the proceedings nor the scholarly papers from this conference were ever published in a single volume. The conference was organized by Costa Carras. Participants at this inter-religious and interdisciplinary gathering included Bill Reilly, who in the following year was appointed director of the Environmental Protection Agency in the United States.

4. See Chapter 1, "Call to Vigilance and Prayer."

as a day for all Orthodox Christians within the jurisdiction of the Ecumenical Patriarchate to offer prayers for the protection and preservation of the natural creation of God. Since that time, a similar statement and spiritual reminder is issued every year, on the first day of September, to Orthodox faithful in North and South America, Western Europe and Great Britain, as well as in Australasia.[5]

In the following year (1990), the foremost hymnographer of the monastic republic on Mt. Athos, Monk Gerasimos Mikrayiannanites (d. 1991),[6] was commissioned by the Ecumenical Patriarchate to compose a service of supplication, with prayers for the protection of the environment.[7] The Orthodox Church has always and traditionally prayed for the environment. Whereas, however, in the past, Orthodox faithful prayed to be delivered from natural calamities, from June 6, 1989, the Ecumenical Patriarch was calling on Orthodox Christians to pray for the environment to be delivered from the abusive acts of the human inhabitants of this planet.[8] Therefore, most of the prayers composed by Fr. Gerasimos are supplications for repentance, invitations for conversion, and cries of nostalgia for a lost paradise, from which we have been alienated by destroying this world.

---

5. Echoing and citing this appeal by the Ecumenical Patriarchate, and in response to the World Summit on Sustainable Development scheduled for Johannesburg, the central committee of the World Council of Churches passed a resolution (Document GEN 17, 6.2) in its meeting of August 26 through September 3, 2002, urging member churches to mark September 1st each year as a day of prayer for the environment and its sustainability. This recommendation was based on a note from its program committee observing: "Some churches have acted with courage and vision in advocating for a sustainable earth. The call of His All Holiness Bartholomew I, the Ecumenical Patriarch of Constantinople, to Christians around the world to celebrate September 1st as Creation Day so as to pray for the World Summit on Sustainable Development, stands testimony to the commitment of the churches to the earth."

6. The surname denotes the monastic community to which Fr. Gerasimos belonged in the skete of Little St. Anne (or "Mikra Hagia Anna").

7. This service was published in 1991 and translated into English by Fr. Ephrem Lash in England. It was among the last services to be composed by Fr. Gerasimos. The translation appeared as *Office of Vespers for the Preservation of Creation* in *Orthodoxy and Ecology: Resource Book* (Bialystok: Syndesmos, 1996). A second English translation of the Vesperal Service has also appeared in the United States by Narthex Press, entitled *Vespers for the Protection of the Environment* (Northridge, CA, 2001).

8. See Ecumenical Patriarchate assisted by the World Wide Fund for Nature, ed., *Orthodoxy and the Ecological Crisis* (Helsinki: WWF International, 1990).

Only one month after rising to the ecclesiastical throne of Constantinople, in November of 1991, the Ecumenical Patriarch initiated and convened an ecological meeting on the island of Crete. The title of the gathering was "Living in the Creation of the Lord." That convention was attended and officially opened by Prince Philip, the Duke of Edinburgh and International Chairman of the World Wildlife Fund.

In the following year, March of 1992, Bartholomew called a meeting of all Orthodox Patriarchs and Primates, who gathered at the Phanar for a historical expression of unity in theological vision and pastoral concern. Here, the Ecumenical Patriarch again introduced the topic of the protection of the natural environment and asked from all the leaders of the Orthodox Churches to inform their churches about the critical significance of this issue for our times. The official message of the Orthodox Primates, representing all of the Orthodox Churches throughout the world, acknowledged, approved, and endorsed the Patriarchal initiative to establish September 1st as a day of Pan-Orthodox prayer for the environment.

In the summer of the same year, the Duke of Edinburgh also accepted to visit the Phanar and address an environmental seminar at the Theological School of Halki. In November of 1993, the Ecumenical Patriarch returned the visit to the Duke, meeting with him at Buckingham Palace where they sealed a friendship of common purpose and active cooperation for the preservation of the environment. From that same year, annual encyclicals of September 1st always emphasize the importance of ecological concern and action.

In April of 1994, the Ecumenical Patriarch was invited to the administrative offices of the European Commission, where he delivered a speech with a significant message. It was the first time that someone who was not a state or political leader had been asked to address the European Commission. The influence and impact of the young Patriarch was broadening to secular and governmental levels.

In June of 1994, another ecological seminar was convened at the historical Theological School of Halki. This was the first of five successive annual summer seminars held at Halki on diverse and original aspects of the environment. The topic was "Environment and Religious Education" (June 20–29, 1994). A second summer seminar, in 1995, focused on the theme "Environment and Ethics" (June 12–18, 1995). The third summer seminar was entitled "Environment and Communications" (July 1–7, 1996); the fourth meeting explored the topic of "Environment and Justice" (June 25–30, 1997); and the fifth gathering approached the subject of "Environment and Poverty" (June 14–20, 1998).

These seminars, the first of their kind, were designed to promote environmental awareness and action inspired by the initiatives of the Ecumenical Patriarchate.[9] They sought to engage leading theologians, environmentalists, scientists, civil servants, and other experts. Speakers have included Church leaders, governmental authorities, scientists and ethicists, academicians and intellectuals, artists and journalists, as well as pioneers of ecological programs worldwide. Participants varied in number, from fifty to eighty, and arrived from all over the world, representing the major Christian denominations and world religions.

In October of 1994, the Department of Environmental Studies of the University of the Aegean conferred an honorary doctoral degree on Patriarch Bartholomew. This was the first of a series of awards and honorary degrees presented to Bartholomew in recognition of his efforts and initiatives to preserve the environment.[10] The Department of Forestry and Environmental Studies of the Faculty of Earth Sciences at the University of Thessalonika bestowed a similar honor on the Patriarch in 1997. In November 2000, the New York–based organization Scenic Hudson, one of the earliest and most prestigious environmental groups in the United States, presented the Ecumenical Patriarch with the first international Visionary Award for Environmental Achievement.

In June 2002, Patriarch Bartholomew was the recipient of the Sophie Prize, the most celebrated environmental award in the world. Established by Jostein Gaarder, the well-known author of *Sophie's World*, the Sophie Prize is presented to an individual or organization that has, in a pioneering and creative manner, pointed to and developed alternative approaches

---

9. See *The Environment and Religious Education* (Istanbul: Melitos Editions, 1995), and *The Environment and Ethics* (Istanbul: Melitos Editions, 1996).

10. A further recognition of the Patriarch's inspiration and vision was the 2002 environmental award of the Binding Institute of Liechtenstein in Switzerland (November 2002). In 2004, he was also named among the United Nations' "Champions of the Earth," and in 2010, he received the Hollister Award from the Temple of Understanding in New York.

to environmental awareness and action. In the case of Patriarch Bartholomew, the award was presented by the Minister of Environment in Norway during an official ceremony in Oslo. The reasons cited for the award were:

His pioneering efforts in linking faith to the environment, reminding all people of faith of their direct responsibility to protect the earth.

His spiritual and practical environmental leadership, managing to raise the environmental awareness of 300 million faithful of the Orthodox Church worldwide and challenging religious leaders of all faiths to do the same.

His tireless efforts to bring attention to both rights and obligations, criticizing both the over-consumption in the first-world countries and the lack of justice that causes growing inequity in developing countries.

The 1999 recipient, Thomas Kocherry of India, made the following remarks about the selection of the Patriarch:

As a religious leader, Patriarch Bartholomew gives meaning to Jesus Christ even today. Through his work, Jesus is still alive and risen. Environmental awareness and social justice go together. His All Holiness has taught the whole world that an institutional church has relevance.

The 2001 recipient, Bernard Cassen of ATTAC in France, observed:

The efforts of His All Holiness in raising environmental awareness . . . as well as exposing the negative impacts of globalization on the poor are examples for other world religious leaders to follow.

In 1994, convinced that any appreciation of the environmental concerns of our times must occur in dialogue with other Christian confessions, other religious faiths, as well as other scientific disciplines, Patriarch Bartholomew established the *Religious and Scientific Committee*. As we share the earth, so too do we share the responsibility for our pollution of the earth and the obligation to find tangible ways of healing the natural environment. This ecumenical and interdisciplinary committee is chaired by the Most Reverend Metropolitan John (Zizioulas) of Pergamon, professor of theology at King's College (University of London), and the Theological School of the University of Thessalonika.

The Religious and Scientific Committee—coordinated over the years by Ms. Maria Becket—has convened eight international, interdisciplinary and inter-religious symposia in order to study and reflect on the fate of the rivers and seas, which cover two-thirds of the earth's surface. These symposia have gathered scientists, environmentalists, journalists, policy-makers, and representatives of the world's main religious faiths in an effort to draw global attention to the plight of the crucial bodies of water of local, regional, and global significance—such as the Mediterranean and Black Seas as well as the Amazon and Mississippi Rivers. Participants met in plenary, workshop, and briefing sessions, hearing a variety of speakers on various environmental and ethical themes. Delegates also visit key environmental sites in the particular region of the symposium.

Symposium I: Revelation and the Environment was convened in September 1995 (September 20–27), under the joint auspices of Patriarch Bartholomew and Prince Philip on the occasion of the 1900th anniversary of St. John's Book of Revelation. This scriptural book portrays the destructive impact of humanity on the earth and the seas with vivid language that has fascinated readers through the centuries for scholarly, scientific, symbolical, spiritual, and sensational reasons.[11]

Traveling on ship through the Aegean and the Eastern Mediterranean, the two hundred participants of this symposium identified the pollution of the world's waters as a threat to the survival of the planet and recommended the creation of a common language for scientific and theological thought to overcome centuries of estrangement and misunderstanding between science and faith. It is so critical to the entire vision of the Patriarch that the first symposium began with an emphasis on and an interpretation of the Book of Revelation, the closing book of the New Testament. Although, in many people's minds, apocalypse would seem to imply destruction and holocaust—which is clearly one aspect of the concept—nevertheless only a literalist and fundamentalist would accept this as the complete interpretation of the term. In fact, "revelation" signifies a vision that evil can be conquered by good, that ugliness can be overcome by beauty.

---

11. See the official publication of the symposium papers: S. Hobson and J. Lubchenko, eds., *Revelation and the Environment AD 95–1995* (Singapore: World Scientific, 1997).

Symposium II: The Black Sea in Crisis was held September 20–27, 1997, under the joint auspices of the Ecumenical Patriarch and His Excellency Jacques Santer, President of the European Commission. This time, the symposium undertook a concrete case study, visiting the countries that surround the Black Sea and engaging in conversation with local religious leaders and environmental activists, as well as regional scientists and politicians.[12]

A direct result of this second symposium, the *Halki Ecological Institute*, was held June 5–20, 1999, in order to promote and provide wider regional collaboration and education among some seventy-five clergy and theologians, educators, and students, as well as scientists and journalists. This educational initiative marked a new direction in the interdisciplinary vision and dialogue concerning the environment, seeking to implement the principles of the ecological vision determined by the Religious and Scientific Committee by turning theory into practice.

The inaugural session of the Institute was conducted in a forested lakeland area outside of Istanbul and then on the island of Halki in the Sea of Marmara. It was attended by seventy members of the clergy, educational institutions, and media in the Black Sea region. Participants received intensive instruction from a team of international and regional theologians, scientists, educators, and environmental policy professionals. The two-week program included theological and scientific presentations in conjunction with field trips and scientific studies of the specific issues faced in that region. As an adjunct to this program, the Institute simultaneously conducted the Black Sea Environmental Journalists Workshop for around twenty print, radio, and television journalists from the six Black Sea nations. The purpose was to provide a network for collaborative efforts in the region. The journalists joined the educational workshop for a series of interdisciplinary sessions on topics of mutual interest and concern.

During the second symposium, it became evident that no solution to the ecological collapse of the Black Sea could be determined without addressing the degradation of the rivers that flow into that sea. Therefore, *Symposium III: "River of Life"—Down the Danube to the Black Sea* was

---

12. See the official publication of the symposium papers: S. Hobson and L. Mee, eds., *Religion, Science, and the Environment: The Black Sea in Crisis* (Singapore: World Scientific, 1998).

launched in October 1999, under the joint auspices of Patriarch Bartholomew and His Excellency Romano Prodi, President of the European Commission. This meeting gathered international and local leaders in the fields
of science, religion, and environmental policy, who traveled the length of
the Danube River, from Passau, Germany to the delta of the Black Sea.
In the aftermath of the military and ethnic conflict in the Former Republic of Yugoslavia, the challenge of protecting and restoring the state of the
waters and natural environment along the Danube River became all the
more critical and urgent. Symposium III focused on the ecological impact
of war, urban development, industrialization, shipping, and agriculture.[13]

Symposium IV: The Adriatic Sea—a Sea at Risk, a Unity of Purpose
addressed the ethical aspects of the environmental crisis. Held in June
2002, under the joint auspices of the Ecumenical Patriarch and His Excellency Romano Prodi, President of the European Commission, this symposium opened in Durres, Albania and concluded in Venice, Italy. The
emphasis during this meeting was on the need to cultivate particular ecological principles and values among peoples in affluent countries and advanced economies as well as among peoples of recovering countries and
transitional economies.[14]

Two unique events during this symposium turned more than a new
page in the "book of nature," sealing them as memorable occasions in the
"book of history." On June 9, 2002, the Ecumenical Patriarch celebrated
the Divine Liturgy of St. John Chrysostom in the church of Sant' Apollinare in Classe for the first time in twelve centuries. Perhaps the most
impressive basilica of early Christianity, this church boasts the tomb of St.
Apollinare (first bishop of Ravenna) as well as unparalleled sixth-century
mosaics.

The spiritual reality expressed in these mosaics suggests a sense of eternity and wonder; it is literally a lifting and entry into heaven. The artistic
decorations display great originality and innovativeness, depicting the
transformation of earth and the correspondence between earth and

---

13. See the official publication of the symposium papers: N. Ascherson and S. Hobson,
eds., *Danube: River of Life* (Athens: Religion, Science and the Environment, 2002).

14. See the official publication of the symposium papers: N. Ascherson and A. Mar-
shall, eds., *The Adriatic Sea, a sea at risk, a unity of purpose* (Athens: Religion, Science and
the Environment, 2003).

heaven. In a way never before represented, the Transfiguration of Christ on Mt. Tabor is rendered in a completely symbolic manner, with the figure of Christ shown in the form of a Cross and the face of Christ at its center, with the voice of the Father replaced by a hand coming out of a cloud and the Apostles shown as lambs ordered around the Cross (further symbols of sacrifice and martyrdom), and with the entire world shown as beautiful rocks, bushes, and flowers. The overwhelming notion is that of reconciliation between Christ, humanity, and all of creation. The connection is also clear between transfiguration and crucifixion, as well as between eucharistic mystery and cosmic resurrection.

The next day, June 10, 2002, delegates attended the closing ceremony in the Palazzo Ducale, where yet another historical moment of ecumenical and environmental significance unfolded. There, Bartholomew co-signed a document of environmental ethics with Pope John Paul II, whose presence was communicated via satellite link-up. The Venice Declaration was the first text ever signed jointly by the two religious leaders on ecological issues and emphasizes that the protection of the environment is the moral and spiritual duty of all people.

Symposium V: The Baltic Sea—A Common Heritage, A Shared Responsibility was organized in June 2003, moving from Gdansk, through Kaliningrad, Tallinn, and Helsinki, and concluding in Stockholm.[15] The Baltic Sea borders on and is polluted by nine countries with widely disparate resources and economies, as well as social structures and religious faiths. The end of the Cold War has permitted the renewal of political, economic, social, cultural, and religious ties between this region and countries making up the European Union, and the wider world. Organized under the patronage of the Ecumenical Patriarch and HE Romano Prodi, President of the European Commission, the symposium sought to draw lessons from the Baltic—its diversity, problems, and history—in order to illustrate the challenges faced by humanity in that region and more widely. The direct result of this symposium was the North Sea Conference, cosponsored by the Ecumenical Patriarchate and the Church of Norway.

---

15. See the official publication of the symposium papers: N. Ascherson and B. Cathcart, eds., *A Common Heritage, A Shared Responsibility: Symposium V* (Athens: Religion, Science and the Environment, 2010).

Symposium VI: The Amazon—Source of Life was held in July 2006 on the Amazon River under the patronage of the Ecumenical Patriarch and His Excellency Kofi Annan, Secretary-General of the United Nations. Participants journeyed from Manaus, through Santarem and Jau, meeting at the crossing of the Rio Negro and the Rio Solimoes for a special ceremony of the blessing of waters by the Ecumenical Patriarch and indigenous leaders. Through scholarly reports, religious papers, and several field trips to the Jau Reserve, Mamiraua, and Lake Piranha, and along the Amazon itself, participants received a unique perspective on the environmental problems of the region. This symposium concentrated on the global dimension of problems stemming directly from the Amazon, problems which had, perhaps, dropped out of view for many decision-makers.

Symposium VII: The Arctic—Mirror of Life, held in Greenland during the fall of 2007, directed its attention to the Arctic Sea and the imminent dangers of global warming.[16] Under the joint patronage of His All Holiness the Ecumenical Patriarch and Their Excellencies Jose Barroso (President of the European Commission) and Kofi Annan (former Secretary-General of the United Nations), *The Arctic—Mirror of Life* considered the suffering of the indigenous populations, the fragility of the sea ice, and the encroachment of oil exploration in a region considered to be one of the first victims of human-induced climate change. Delegates visited areas where the impact of melting ice is already clear, the northernmost communities in the world, which have demonstrated extraordinary resilience in the face of change, as well as the towering edge of the ice mass, still vast albeit retreating yearly toward the Pole. There, the assembled religious leaders of various faiths and disciplines joined in prayer for the protection and preservation of the planet.

Symposium VIII: Restoring Balance—The Great Mississippi River was held in New Orleans in October 2009 and considered the immediate impact of Hurricane Katrina, the devastating consequences of regional development, and the overall influence of global corporate capitalism on the Gulf region as well as on climate change in general. Throughout this symposium, the moral aspects of the ecological problem were examined, particularly with respect to the religious and cultural background of the

---

16. For a statement by the Patriarch on global warming, see http://tiny.cc/diakonia.

United States of America, so that religious communities can cooperate in facing the ecological crisis and learn from the Mississippi River in order to respond to challenges of other major river and delta systems around the world.

What became evident during the symposium was that, as in similar cases elsewhere throughout the world, the release of toxic materials in and along the Mississippi River poses a variety of threats to human beings and to the natural environment. In most cases, it is the poor and vulnerable who are most affected by contaminants released into the air, water, and land. The tension between the economic benefits of petrochemical production and the rights of those living nearby represents one of the fundamental difficulties in achieving what has become known as environmental justice. This symposium proved prophetic in its critical speeches and statements, especially in light of the tragic oil spill in the Gulf less than six months later, on April 20, 2010, as a result of British Petroleum's Deepwater Horizon off-shore drilling.

READING THIS VOLUME

The collection of encyclical letters and addresses opens with a call to prayer. This entire first section includes the letters written and published by the Ecumenical Patriarch on the occasion of the beginning of the ecclesiastical year, which in the Orthodox Church is commemorated and celebrated as the feast of the *Indictus* on September 1st. The first two encyclicals (1989 and 1990) were signed by the hand of the late Patriarch Dimitrios (d. 1991).[17] However, they are an important part of this collection for two reasons: historically, because they mark the initial steps of the Ecumenical Patriarchate toward educating and mobilizing its adherents on environmental issues; and personally, because Patriarch Bartholomew was the closest coworker and advisor of Patriarch Dimitrios. Although Bartholomew himself will always and formally—out of a sense of humility, but especially out of a reverence for tradition—acknowledge Patriarch Dimitrios, his "revered predecessor" as he will often describe him, recognizing him as the person responsible for the call to prayer for the preservation and protection of the natural environment, in fact he has

---

17. See Chapter 1 of this book, "Call to Vigilance and Prayer."

at all times been either behind the scenes or else at the forefront of this unique and pioneering ministry.

Thereafter, encyclicals are written and signed by Patriarch Bartholomew every year, with the exception of the years 1991 and 2000. September 1991 marked the death of the late Dimitrios, as well as the installation of Bartholomew himself; no encyclical was circulated during that year. September 2000 marked the special millennial celebrations of the Ecumenical Patriarchate, and once again no special encyclical appeared during that jubilee. Nevertheless, even in those two years, environmental issues were not too far from the concern of the Patriarch. In 2000, the Patriarch addressed the young people of his jurisdiction throughout the world and emphasized critical contemporary issues, such as the environment and unemployment. Moreover, from as early as 1991, Bartholomew addressed the entire Church with his formal enthronement address, which contains the seeds of the fundamental theological and spiritual principles that guide his ecological vision.

From the outset of his tenure, the young Patriarch Bartholomew is able to see the larger picture. He recognizes that he stands before something greater than himself; he senses that he belongs to an "unbroken" succession or chain that long predated him and would long outlast him. He is a part of this tradition, and cannot conceive himself apart from this same tradition. Therefore, he speaks of self-emptying (or *kenosis*), of ministry (or *diakonia*), of witness (or *martyria*, a Greek term that also has the sense of martyrdom and suffering!),[18] and of thanksgiving (or *eucharistia*, a Greek term that also implies liturgy!). Indeed, Bartholomew repeatedly makes mention of his own spiritual elders, frequently referring especially to his predecessor, the late Patriarch Dimitrios.

The same signature of humility is evident on other levels as well. For instance, Bartholomew always describes the Ecumenical Patriarchate as a spiritual institution; it is not, he notes, a powerful establishment in secular terms. The same point will be underlined elsewhere, such as in his address

---

18. Martyrdom is also an important feature of Bartholomew's addresses to the Primates of the Orthodox Churches worldwide (1992 and 1995). The concept of sacrifice is the essential—described as a "missing"—dimension of environmental ethics of the closing address during the fourth international and inter-religious symposium (2002).

in Kathmandu (2000). The emphasis instead is on simplicity—the technical term here in Orthodox theology is *asceticism* (or the ascetic life).[19] In this respect, specific reference is made to the monastic tradition and particularly to the existence of monks on Mt. Athos for over one thousand years. The emphasis in the enthronement address is also on *liturgy*, which is the source and essence of Orthodox theology and spirituality.[20]

The notion of liturgy leads us into what is perhaps the most distinctive characteristic of Patriarch Bartholomew's vision, already foreshadowed from the day of his enthronement, namely the central and crucial concept of *communion*.[21] In everything that Bartholomew says and does, particularly in light of the environmental crisis, he is aware that all of the Orthodox must be included. Indeed, not only should all Orthodox Christians be in communion, but also all Christians in general should be in communication. In addition, all religions should be in cooperation;[22] all sciences and disciplines should be committed;[23] all cultures and ages should concur; and even atheists should be seen to contribute in the movement toward the heavenly kingdom. In his welcome address to the third summer seminar on Halki in June 1996, Patriarch Bartholomew noted:

---

19. This emphasis is found in several texts, such as the opening address at the Black Sea symposium (1997), the address in Santa Barbara (1997), the meditation on the Cross (1999), the closing address on the Adriatic symposium (2002), the address at Utstein Monastery (2003), in the Patriarch's article for the *International Journal of Heritage Studies* (2006), as well as in the encyclicals for September 1st in 1994 and 1997; in the latter encyclical, Patriarch Bartholomew states that the ascetic way is the "original commandment" from the moment of the Genesis creation of humanity.

20. The emphasis on liturgy may be found in several of the addresses during the Patmos symposium (1995), in the opening of the Amazon symposium (2006), as well as in the various encyclical for September 1st.

21. See A. Belopopsky and D. Oikonomou, eds., *Orthodoxy and Ecology: Resource Book* (Athens: Syndesmos, 1996); and Ecumenical Patriarchate (with Syndesmos), *So That God's Creation Might Live: The Orthodox Church Responds to the Ecological Crisis*. Proceedings of the Inter-Orthodox Conference on Environmental Protection, Crete, 1991.

22. See his opening address during the Patmos symposium (1995) and the Patriarch's message for World Oceans Day (2003).

23. See especially the opening of the Patmos symposium (1995), the opening address of the Black Sea symposium (1997), and the closing address for the Amazon symposium (2006).

Unhindered communication among all those concerned with the management of the ecological realities of the present time is equivalent to the indispensability and the sanctity of prayer.

Patriarch Bartholomew always defers to others: to God and the saints, to his teachers and predecessors, to his colleagues and contemporaries, to scientific knowledge. The overwhelming sense of humility is, therefore, extended not only upward (before the grace of God) and backward in time (before the face of tradition), but also sideward (before every human face) and downward (before the beauty of the natural world). In other texts, Patriarch Bartholomew will speak of humility before the earth and even of dialogue with nature.[24] This is the way of his Church. It is the way of authenticity. And it is what provides him with an overriding sense also of authority.

It is for this reason that Patriarch Bartholomew considers his prayer for and protection of the environment as an obligation, not as a way of submitting to contemporary fashions or political statements. His commitment to and involvement in environmental issues are not a matter of public relations but of theological conviction. This is very evident in his interviews, found in the final section of this book. It is also apparent in his repeated phrase, "We cannot remain idle!"[25] He makes it clear that he will not cease proclaiming the importance of the environment.[26] It is, he believes, an almost "apostolic commission."[27]

For Patriarch Bartholomew, it is a matter of truthfulness to God, humanity, and the created order. In fact, it is not too far-fetched to speak of environmental damage as being a contemporary heresy or natural terrorism; he condemns it as nothing less than sin![28] The environment is not a political or a technological issue; it is, as the Patriarch underlines in so

---

24. See his addresses at the Halki Ecological Institute (1999) and in Novi Sad (1999). In the encyclical for September 1st (2001), Patriarch Bartholomew underlines the inherent danger in human arrogance over nature.

25. Encyclical by Patriarch Dimitrios for September 1st (1989) and address by Patriarch Bartholomew during the Patmos symposium (1995). See also his homily at Uspenski Cathedral, Helsinki (2003).

26. See the Patriarch's address at the closing of the Danube symposium (1999).

27. See the Patriarch's opening address at the Black Sea symposium (1997).

28. See his address at the University of the Aegean (1994), the encyclical for September 1st (1996), the address at Santa Barbara (1997), and the interviews.

many of his interviews, primarily a religious and spiritual issue. Religion, then, has a key role to play; and a spirituality that is not involved with outward creation is not involved with the inward mystery either.[29]

Once again, all of this accounts for Patriarch Bartholomew's sense of authority. He offers no apologies for his traditional Orthodox theological background and starting point. He offers no apologies for his criticism of a-religious responses to the environment, whether these derive from scientific or moral sources.[30] In addition, he offers no apologies for his harsh criticism of false technological promises about "progress" and "development."[31] On the other hand, he is quite open about the thirst of western civilization for another worldview and spirituality.

Moreover, he is not afraid to adopt "politically incorrect" language concerning the centrality of the human person within creation.[32] It is not anthropocentrism that is the problem, he would feel; it is *anthropomonism*,[33] namely the exclusive emphasis on and isolation of humanity at the expense and detriment of the natural environment. Nature is related to people and people to nature.[34] If Bartholomew is radical in the theological articulation of his vision, it is because his worldview is rooted (the literal sense of the word "radical") in his Church tradition.

All of this leads Bartholomew to another dimension of his environmental initiatives, namely the deeply *pastoral attitude* toward those whom he addresses. Whether he is speaking to his colleagues among the worldwide Orthodox Primates (as in his address in 1992), or to individual leaders and churches (as in Galati in 1999), or else to the Orthodox faithful throughout the world (as in all of the encyclicals for September 1st), Ecumenical Patriarch Bartholomew is sensitively, thoughtfully, and gradually

---

29. On the role of religion, see the Patriarch's addresses on the Danube (Bulgaria, 1999); on spirituality, see his encyclical for September 1st (1994).

30. Encyclicals for September 1st (1992 and 1996), his words at the University of the Aegean (1994) and at the Halki seminar (1995), and his address in Atami, Japan (1995).

31. Encyclical for September 1st (1994). See also his address at the Lutheran Church of Stockholm (2003).

32. See the Patriarch's encyclicals for September 1st (1993, 1995, 1997, 1998, 2001) and his addresses at the opening of the Black Sea symposium (1997), in Japan (1998), on Halki (1998), and at the opening of the Danube symposium (1999).

33. See especially the interviews that appear in Chapter 8 of this book.

34. See especially the encyclical for September 1st (2001) and the interviews that appear in Chapter 8 of this book.

educating with his "paternal admonition."[35] He recognizes that, in many ways, although he is firmly rooted in the vine of his Church, he is also "out on a limb." He constantly needs to remind his peers and his spiritual children. This is precisely why, as a patient pedagogue, he does not tire in his efforts, even when they appear slow to bear fruit.

Readers will notice an air of repetition, even repetitiveness, in many of the texts that follow. Bartholomew will frequently refer—sometimes in the same phraseology—to his vision, the Patriarchate's initiatives, and the Orthodox theological reasons for these. For instance, there is repeated mention of the encyclicals for September 1st, the Summer Seminars on Halki, and the various sea-borne international symposia. This is not a sign of arrogance, but yet another aspect of humility. Bartholomew knows that he must continually and vigilantly inform his people. I have personally witnessed the way in which, particularly during the international symposia, Patriarch Bartholomew will seize every occasion to visit local Orthodox leaders, gently but firmly eliciting their understanding and support in the concern for the natural environment. Indeed, the international symposia have had a practical and symbolical value for the visible unity of the Orthodox Church, inasmuch as they have afforded the Ecumenical Patriarch the opportunity of visiting numerous Orthodox Churches in the regions.

Patriarch Bartholomew will invariably relate the environment to a familiar aspect of Orthodox spirituality, namely to the *icons* that decorate Orthodox churches. Symbols are important in Orthodox thought, worship, and life.[36] Creation itself is likened to an icon,[37] in the same way as the human person too is created "in the image [or, icon] and likeness of God" (Gen. 1.26). And he invites the Orthodox to contemplate the Creator God through the icon of the created world. Creation is a visible and tangible revelation [or *apocalypse*] of the presence [or *parousia*] of the Word of God.[38] Humanity is called to wonder at creation, but not to worship creation. Otherwise, the natural world is reduced from the level

---

35. Encyclical for September 1st (1996).

36. See the opening address during the Patmos symposium (1995), and the Patriarch's address at the Institute of Marine Research (2003).

37. See the Patriarch's address in Santa Barbara (1997).

38. Apocalypse is perceived both as the end (as in the encyclical for September 1st in 1989) and as the epiphany of God (as in the opening address during the Patmos symposium in 1995).

of icon to the level of idol.[39] It is in the same vein that Patriarch Bartholomew will often refer to the human beings as made and intended by God to serve as "priests" within the created world.[40]

Patriarch Bartholomew believes that a particular ethos is called for in our response to the environmental crisis.[41] This ethos is divinely inspired, even commanded, by God![42] And it is an ethos incarnated, even communicated, by the Saints![43] The Saints are those who inform and influence, who re-form and instruct the world through their presence and their prayer. They remind us, says Bartholomew, that the personal responsibility and the slightest action of even the feeblest among us can change the world for the better.[44]

Finally, the Ecumenical Patriarch is aware of the reality that environmental issues are intimately connected to and dependent on numerous other social issues of our times[45] and our ways, to which he will repeatedly refer in his addresses and remarks. These include such subjects as war and peace,[46] justice and human rights,[47] unemployment and

---

39. See the opening address on the Black Sea (1997) and the encyclical for September 1st (1999).

40. See, for instance, his address at the Halki seminar (1992) and the foreword to Symposium V proceedings (2003).

41. This ethos is well articulated by the leading spokesman for the ecological initiatives of the Ecumenical Patriarchate, Metropolitan John (Zizioulas) of Pergamon. See his "Preserving God's Creation: three lectures on theology and ecology," *King's Theological Review*, xii (1989) and also "Ecological Asceticism: a cultural revolution," *Our Planet*, vii, 6 (1995).

42. See his addresses at the University of the Aegean (1994) and at the summer seminar on Halki (1995), and his homily in Santa Barbara (1997), but especially the encyclical for September 1st (1997).

43. Encyclicals for September 1st (1997, 1998, and 1999).

44. See especially the encyclical for September 1st (1998).

45. See the first volume in this series, entitled *In the World, Yet Not of the World: Social and Global Initiatives of Ecumenical Patriarch Bartholomew* (New York: Fordham University Press, 2010).

46. See his addresses at the opening of the Patmos symposium (1995), the closing of the Black Sea symposium (1997), the opening of the Danube symposium, as well as during its journey through Bulgaria (1999) and in Novi Sad (1999). See also the encyclical for September 1st (1998).

47. See the foreword for the summer seminar (1997), as well as his address during the closing of the Danube symposium (1999).

poverty.[48] Even on occasion when the environment does not appear to be the central issue of a particular address, nevertheless the interconnectedness of the above-mentioned problems with the destruction of our planet becomes profoundly apparent to the discerning reader. It is not by chance that the term "eco-justice" has been coined in ecclesiastical circles to describe this interconnection between creation and creatures, between the world and its inhabitants. We have, in recent years, become abundantly aware of the effects of environmental degradation on people, and especially the poor.

Everything that the Patriarch says and does is colored by his keen awareness of and deep concern for environmental issues. There is a sense in which—based on the spiritual tradition of icons, liturgy, and asceticism—the environment is implied in the very silence and prayer of the Orthodox Church. For Ecumenical Patriarch Bartholomew, the created world is part and parcel of our salvation; we shall—and can only be—saved with the environment.[49] Therefore, ecology is very much at the center of his thinking and his action; it constitutes the central plate on the altar of his worship, on the desk of his writing, as well as on the agenda of his ministry.

---

48. See his encyclical for September 1st (1994), the foreword for the summer seminar (1998), and his address in Halki (1998). Especially, see his statement to the World Council of Churches (2004).

49. For Orthodox theological monographs on the environment, see J. Chryssavgis, *Beyond the Shattered Image: Orthodox Insights into the Environment* (Minneapolis: Light and Life Books, 1999); E. Theokritoff, *Living in God's Creation: Orthodox Perspectives on Ecology* (Crestwood, NY: St. Vladimir's Seminary Press, 2009); P. Sherrard, *The Eclipse of Man and Nature: An Enquiry into the Origins and Consequences of Modern Science* (Felton, Northumberland UK: Lindisfarne Press, 1987); P. Sherrard, *Human Image, World Image* (Ipswich UK: Golgonooza Press, 1990); Kallistos Ware, *Through the Creation to the Creator* (London: Friends of the Centre Papers, 1997); G. Limouris, ed., *Justice, Peace, and the Integrity of Creation: Insights from Orthodoxy* (Geneva: WCC Publications, 1990); and Paulos Gregorios, *The Human Presence: An Orthodox View of Nature*, Geneva: World Council of Churches, 1987 (later published as *The Human Presence: Ecological Spirituality and the Age of the Spirit*).

1

Call to Vigilance and Prayer

Patriarchal Encyclicals for September 1st

MESSAGES BY ECUMENICAL PATRIARCH DIMITRIOS
(1972–1991)

DIMITRIOS
By the Mercy of God,
Archbishop of Constantinople–New Rome and Ecumenical
  Patriarch,
To all the Faithful of the Church:
Grace, mercy, and peace from the Creator of all Creation,
Our Lord and Savior Jesus Christ[1]

September 1, 1989
The Church Cannot Remain Idle

This Ecumenical Throne of Orthodoxy—in its responsibility to protect and proclaim the centuries-long spirit of the patristic tradition as well as in its effort faithfully to interpret the eucharistic and liturgical experience of the Orthodox Church—watches with great anxiety the merciless trampling and destruction of the natural environment caused by human beings, with extremely dangerous consequences for the very survival of the natural world created by God.

---

1. The customary address in Patriarchal encyclical letters for September 1, the opening of the ecclesiastical year and the day dedicated to prayer for the protection and preservation of the natural environment.

The abuse by contemporary humanity of its privileged position within creation and of the Creator's order "to have dominion over the earth" (Gen. 1.28) has already led the world to the edge of apocalyptic self-destruction. This is occurring either in the form of natural pollution, which is dangerous for all living beings, or in the form of extinction for many species of the animal and plant world, or else again in various other forms. Scientists and other scholars warn us now of the danger, and speak of phenomena threatening the life of our planet, such as the so-called "green-house phenomenon" whose first indications have already been observed.

In view of this situation, the Church of Christ cannot remain unmoved. It constitutes a fundamental dogma of her faith that the world was created by God the Father, who is confessed in the Creed as "maker of heaven and earth and of all things visible and invisible."[2] According to the great Fathers of the Church, the human person is the prince of creation, endowed with the privilege of freedom. Being a partaker simultaneously of the material and the spiritual world, humanity was created in order to refer creation back to the creator, in order that the world may be saved from decay and death.

COMMUNION AND CONSUMPTION

This great destiny of man was realized, after the failure and fall of the "first Adam," by "the last Adam," the Son and *Logos* of God incarnate, our Lord Jesus Christ, who united in His person the created world with the uncreated God, and who unceasingly presents creation to the Father as an eternal eucharistic referring (Greek: *anaphora*) and offering (Greek: *prosphora*). At each Divine Liturgy, the Church continues this reference and offering (of creation to God) in the form of the bread and the wine, which are elements taken from the material universe. In this way, the Church continuously declares that humanity is destined not to exercise power over creation, as if it were the owner of it, but to act as its steward, cultivating it in love and referring it in thankfulness, respect, and reverence to its Creator.

Unfortunately, in our days under the influence of an extreme rationalism and self-centeredness, humanity has lost the sense of sacredness of

---

2. From the "symbol of faith" or creed articulated in Nicea (325) and Constantinople (381).

creation and acts as its arbitrary ruler and rude violator. Instead of the eucharistic and ascetic spirit with which the Orthodox Church brought up its children for centuries, we observe today a violation of nature for the satisfaction not of basic human needs, but of man's endless and constantly increasing desire and lust, encouraged by the prevailing philosophy of the consumer society.

Yet, "the whole of creation groans and travails" (Rom. 8.22) and is now beginning to protest at its treatment by human beings. Humanity cannot infinitely and whimsically exploit the natural sources of energy. The price of our arrogance will be our self-destruction if the present situation continues.

In full consciousness of our duty and paternal spiritual responsibility, having taken all the above into consideration and having listened to the anguish of modern humanity, we have reached a decision, in common with the Sacred and Holy Synod surrounding us. Accordingly, we declare the first day of September of each year a day on which, on the feast of the Indiction,[3] prayers and supplications are to be offered in this holy center of Orthodoxy for all creation, declaring this day to be the day of the protection of the environment.

Therefore, we invite through this, our Patriarchal Message, the entire Christian world, to offer together with the Mother Great Church of Christ, the Ecumenical Patriarchate, every year on this day, prayers and supplications to the Maker of all, both as thanksgiving for the great gift of creation and in petition for its protection and salvation. At the same time, on the one hand, we paternally urge all the faithful in the world to admonish themselves and their children to respect and protect the natural environment. On the other hand, we urge all those entrusted with the responsibility of governing the nations to act without delay taking all necessary measures for the protection and preservation of the natural creation.

September 1, 1990
Stewards, Not Proprietors

A year has passed since we issued a message declaring September 1st of each year as the day of prayers for the protection of the environment. In

---

3. The first day of the ecclesiastical calendar.

that message, we called upon the Orthodox faithful and, indeed, upon every man and woman of good will, to consider the extent and the seriousness of the problem generated as a result of the senseless abuse of material creation by human beings. On this day, which for the Orthodox Church is officially the first day of the ecclesiastical year, the Orthodox faithful are invited to offer prayers to the Creator of all, thanking Him for the good of His creation and beseeching Him to protect it from every evil and destruction.

The message offered last year continues to echo and to express the inquietude of our Church toward the continuing destruction of the natural environment. For this reason, we considered it expedient, rather than communicating in some other way, to present once again today that same message, unabridged, to our faithful and to all people in order to renew our fervent appeal to all.

Beloved brothers and spiritual children, use the natural environment as its stewards and not as its owners. Acquire an ascetic ethos bearing in mind that everything in the natural world, whether great or small, has its importance for the life of the world, and nothing is useless or contemptible. Regard yourselves as being responsible before God for every creature and treat everything with love and care. Only in this way shall we be able to prevent the threatening destruction of our planet and secure a physical environment where life for the coming generations of humankind will be healthy and happy.

Messages by Ecumenical Patriarch Bartholomew

BARTHOLOMEW
By the Mercy of God,
Archbishop of Constantinople–New Rome and Ecumenical
    Patriarch,
To all the Faithful of the Church:
Grace, mercy, and peace from the Creator of all Creation,
Our Lord and Savior Jesus Christ[4]

---

4. The customary address and introduction of encyclical letters on the occasion of the opening of the ecclesiastical year on September 1, which also marks the prescribed day of prayer for the protection and preservation of the natural environment in the Orthodox Church.

September 1, 1992
Matter and Spirit

The beginning of the ecclesiastical year, sanctified by the traditional celebration of the *Indictus*, also constitutes a characteristic juncture in the life of the entire creation that surrounds us.[5] This juncture is known, at least to those residing in the northern hemisphere of our planet, as the commencement of autumn; to those living in the southern parts of the world, it is known as the beginning of spring.

Thus, "autumn" and "spring," which to the average person usually signify diametrically opposed factors, actually converge and coincide in the inauguration of the ecclesiastical year as *one entity* established by God.

The faithful, therefore, are able to recognize that, in essence, beginning and end constitute two aspects of the same created reality, which is bound to march toward its final destination in both "glory" and "infamy." Therefore, we should not allow the shape and rhythm of the present world to frighten us. In accepting this fundamental truth, we become steadfast and immovable upon the rock of faith. Thus, our sorrowful journey through whatever is "passing" and whatever is "stable" is delivered, at the very outset, from the moral danger, which ever lies in ambush, namely, that of elevating ourselves to the whim of power, which starts from the ground up or else of sinking impiously into the obscurity of despair out of worldly weakness.

In the language of the Church Fathers, the human person "stands at the border" between material and spiritual creation. Humanity is a "borderer" with regard to space and time as well; thus, in "an hour of temptation," one is able courageously to foretaste the "day of salvation."

However, through the sacred correlations mentioned above, creation is by no means at all reduced to a level of irresponsible relativity or relativism. On the contrary, in this way, creation emerges in its God-pleasing uniqueness and sacredness. Thus, "summer" and "winter," "light" and "darkness," "greatness" and "smallness," "instant" and "eternal," "material" and "spiritual," "divine" and "human," are proven not to be contrary to each other, but rather to be deeply correlated. For, the redeeming

---

5. There was no encyclical published in September 1991. Patriarch Dimitrios died on October 2 and Patriarch Bartholomew was elected on October 22.

will of the benevolent God, who is beyond all things, is realized gradually in time and in space through all of these things.

It is, however, precisely within the framework of the sacred connection and correlation of these ideas that God has not allowed humanity to be a mere spectator or an irresponsible consumer of the world and of all that is in the world. Indeed, humanity has been called to assume the task of being primarily a partaker and a sharer in the responsibility for everything in the created world. Endowed, therefore, from the beginning with "the image of God," humanity is called to continuous self-transcendence so that, in responsible synergy with God the Creator, each person might sanctify the entire world, thus becoming a faithful "minister" and "steward."

MINISTER AND STEWARD

It is clear that the concepts of *minister* and *steward*[6] by far exceeds the contemporary and internationally accepted ideal of the person called "an ecologist." Usually, we know neither how such an ecologist understands the concept of *oikos* (the Greek word for "house") nor how that person regards *logos* (the Greek term for "word"). Today everyone speaks of the dangers facing the "ecosystem" as numbering in the thousands, yet few make any reference at all to the God who "constituted" all things. There are those who anxiously keep records of constantly perishing "deposits" of the main elements of life and movements in nature, again without uttering a word about God, who in His infinite goodwill and beneficence is the "depositor" of all His goods for our use and nourishment. In wisdom God "established heaven and earth," thus abundantly enriching the universe with every kind of source of living water.

At any rate, being God's minister and steward over all of creation does not mean that humanity simply prospers or is happy in the world. This would be crude self-sufficiency and impious minimalism. The main and lasting benefit of these qualifications is that, by using the world in a pious manner, humanity experiences the blessed evolution from the stage of "divine image" to that of "divine likeness." In similar fashion, every other good element of the universe is transformed, by the grace of God and

---

6. The Greek word for "steward" is *oikonomos.*

even without human intervention, from the stage of "potentiality" to that of "actuality," in fulfillment of the pre-eternal plan of the entire divine economy.

In addressing these pious thoughts to the faithful of our Church and every person of goodwill, it is our desire, in a manner that is worthy of and pleasing to God, to commemorate and celebrate the inauguration of the *Indictus* as the special day for the protection of all creation. This day was established three years ago by the Mother Holy Great Church of Christ and has now been accepted by all Orthodox throughout the world. Having done so, we should like to take the God-given opportunity to invite and encourage every person, and above all the faithful, to constantly watch over his or her fellow human beings and the world, for the benefit of us all and for the glory of the Creator.

Our words on this auspicious day, and the sacred thoughts, which reach beyond these words, are even more timely inasmuch as they are addressed from the sacred center of the Phanar[7] on the occasion of the first and historic assembly of all the hierarchs in active service of the most venerable Ecumenical Throne. Through this sacred assembly, the Mother Church seeks more direct cooperation and better coordination by the Holy and Sacred Synod with the hierarchs of the Ecumenical Patriarchate throughout the world, namely with those who shepherd dioceses and those who serve in some other capacity. This assembly of our hierarchy, which the Holy Synod decided to convene on a biennial basis, brings numerous blessings, such as closer communication and communion among brothers who share responsibility, as well as an exchange of information and a mutual support. However, it further provides great comfort and encouragement to the children of the Church scattered in the four corners of the earth and represented here by their spiritual leaders. When the faithful around the world from time to time see all of their hierarchy presented and represented as one body, then they recognize it as "divine intervention" against the temptations, sorrows, and dangers in the world, and thus feel greater security in God.

---

7. *Phanar*, meaning "lighthouse" in Greek, refers to the old lighthouse quarter of Istanbul, and it is also the main quarter for Greeks. The name is also seen as coterminous with the Ecumenical Patriarchate since the residence, administrative offices, and cathedral of the Patriarch are there.

September 1, 1993
Creation Ex Nihilo

Together with the other most holy sister Orthodox Churches, we have established September 1st of each year as a special day of concern for and commemoration of the natural environment that surrounds us. Again, this year, we are called to offer wholehearted praise to the Creator of everything both visible and invisible for having placed us as the ones first fashioned in luscious paradise among all His own creation.

The most fundamental Orthodox doctrine—which addresses impartially the omnipotence, the omniscience, the extreme beneficence, and the wakeful providence of the Creator, as well as the consideration and high regard in general for created beings and matter, with humanity as its crowning point—is indeed the doctrine of the creation of the world *ex nihilo* (literally, "out of nothing").

Some people contemplate only what concerns the world and recognize the philosophical "web of the Athenians." So they speak with irony of the conviction of faithful believers with regard to creation *ex nihilo*. In challenging this fundamental doctrine, they cite the merit of the corrupted and, in its redundancy, the frivolous and refutable notion that "nothing can be amassed *ex nihilo*." The only exception such people accept is that, before there was absolute nothingness in relation to the world, God, as being without beginning or successor, and as being beyond and above space, time, quality, quantity, causal relationship, or dependency, has always preceded and commanded everything.

In his epigrammatic statement that "God is love" (1 Jn 4.16), St. John the Evangelist attributed to God, who lacks nothing and is without beginning, a compendious and comprehensive name, that of love, which is cardinal to all moral attributes. Therefore, we, who have received the revealed word of God, are justified in believing that everything has been created out of *absolute love* and in *absolute freedom* by God the Maker and Father of all who, according to St. Paul, calls "things from non-being into being" (Rom. 4.17).

In contemplating the creation of God within us and around us within this kind of God-given theological perspective, we are certainly justified in being overcome with total optimism, even when the elements of nature are faced with the greatest danger or when history is being gravely distorted. For, we recognize that "the souls of the righteous are in the hand of God, and no torment will ever touch them" (Wisd. of Sol. 3.1).

ADHERING TO THE PRECEPTS OF GOD

Therefore, before any abnormality in nature and history, the first requirement is not so much that we be wise and powerful in order to foresee in timely fashion and deal accordingly with certain earthquakes, floundering, or some other usually unexpected calamity. Neither is it that we be armed with the provisions of worldly knowledge and science in order to drive back the powers marshaled against us by any enemy or invader. Rather, above all, we must be just, striving at every moment throughout our life to learn the *precepts* of God more perfectly and more profoundly.

This is why it is not incidental that, among the first things we do in Orthodox worship, is to praise the Lord, invoking Him that we be taught His immovable precepts, which derive from Him only. Never are we so powerful and shielded from every unexpected force, as when we chant, as did the youths in the fire described in the Book of Daniel, the ode of the beloved: "Blessed are you, Lord, teach me your precepts."

During this period, dear brothers and children in the Lord, we witness international organizations, interstate legislations and scientific research programs united in jeremiads and lamentations that toll the bell of danger in order that humanity might sober up in time before the coming of mass chaos. Such chaos, they say, would threaten universal order and balance in the various so-called "ecosystems," not only of our planet, but also of the entire cosmos. Yet, during this same time, we, from the Ecumenical Patriarchate, address ourselves first to the conscience of every individual person, invite people each day and with innocent heart to taste the good things of God, partaking in trembling fear—though, simultaneously, in doxology and joy—of the good things of creation.

Panic has never allowed humanity to render judgments calmly or to balance justly its obligations toward itself, toward the world around, and toward the ever-watchful God above. However, it is precisely these obligations, as they have been coordinated from the very first moment of creation by the just-judging God, which constitute the "precepts" mentioned above. Usually, people speak out and go to great pains to mark and establish *human rights*, which, as a rule, are determined by self-interest and fear and always give rise to powers and demands, which separate persons from groups, from classes, from people.

The precepts of God, on the contrary, are by definition comprehensive and inclusive, as much of *the part* as of a *whole*. This is why, by learning and recognizing them, human beings are rendered, through God's grace, brothers and partakers among themselves. Furthermore, through a eucharistic use of the world, they are rendered partakers of the world and of the infinite love of God, and not consumers, which the atheistic polity or eudemonistic instinct has taught through the hubristic progress of technology.

Thus, the first responsibility of the faithful is to examine and study continuously in greater depth the law and precepts of God. Thus, by becoming cheerful givers and grateful receivers of His wondrous things in this world, we may come to respect the balances of nature set up by Him.

September 1, 1994
All of Creation Groans

On a number of occasions in the ecclesiastical year, the Church prays that God may protect humanity from natural catastrophes: from earthquakes, storms, famine and floods. Yet, today, we observe the reverse. On September 1st, the day devoted to God's handiwork, the Church implores the Creator to protect nature from calamities of human origin, calamities such as pollution, war, exploitation, waste, and secularism. It may seem strangely paradoxical that the body of believers, acting vicariously for the natural environment, beseeches God for projection against itself, against its own actions. Nevertheless, from this perspective, the Church, in its wisdom, brings before our eyes a message of deep significance, one that touches upon the central problems of fallen humanity and its restoration. This is the problem of the polarization of individual sin against collective responsibility.

Scripture informs us that if one member of the body is infirm, the entire body is also affected (1 Cor. 12.26). There is, after all, solidarity in the human race because, made as they are in the image of the Trinitarian God, human beings are interdependent and co-inherent. No man is an island. We are all "members one of another" (Eph. 4.25). Therefore, any action, performed by any member of the human race, inevitably affects all other members. Consequently, no one falls alone and no one is saved alone. According to Dostoevsky's Staretz Zosima in *The Brothers Karamazov*, we are each of us responsible for everyone and everything.

How does this central problem relate to the matter of protecting the environment against humankind's actions? It has become painfully apparent that humanity, both individually and collectively, no longer perceives the natural order as a sign and a sacrament of God but rather as an object of exploitation. There is no one that is not guilty of disrespecting nature. To respect nature is to recognize that all creatures and objects have a unique place in God's creation. When we become sensitive to God's world around us, we grow more conscious also of God's world within us. In beginning to see nature as a work of God, we begin to see our own place as human beings within nature. The true appreciation of any object is to discover the extraordinary in the ordinary.

Sin alone is mean and trivial, as are most of the products of a fallen and sinful technology. Yet, it is sin that is at the root of the prevailing, destruction of the environment. Humanity has failed in what was its noble vocation: to participate in God's creative action in the world. It has succumbed to a theory of development that values production over human dignity and wealth over human integrity. We see, for example, delicate ecological balances being upset by the uncontrolled destruction of animal and plant life or by a reckless exploitation of natural resources. It cannot be over-emphasized that all of this, even if carried out in the name of progress and well-being, is ultimately to humankind's disadvantage.

It is not without good cause, therefore, that "all of creation groans and travails" (Rom. 8.22). Was it not originally seen by God to be good? Created by God, the world reflects divine wisdom, divine beauty, and divine truth. Everything is from God, everything is permeated with divine energy; in this lies both the joy and tragedy of the world and of life within it. The hymns and prayers in the Office for September 1st, composed by the talented hymnographer of the Great Church, the late monk Gerasimos of the Holy Mountain,[8] extol the beauty of creation but also remind

---

8. Fr. Gerasimos was the elder of a community at Little St. Anne's Skete on Mt. Athos. A refined and renowned hymnographer of the Orthodox Church, he died in 1991.

us of our tragic abuse of it.[9] They call us to repent for our actions against God's gift to us. We have made this world ever more opaque, rendering it ever more tortured. The consequences of nature's confrontation with humanity have indeed been an unnatural disaster of enormous proportions. Is it not, therefore, only right that we Christians act today as nature's voice in raising its plea for salvation before the throne of God?

AN ECOLOGICAL ETHOS

The Church teaches that it is the destiny of humankind to restore the proper relationship between God and the world, just as it was in the Garden of Eden. Through repentance, two landscapes, the one human and the other natural, can become the objects of a caring and creative effort. However, repentance must be accompanied by three soundly focused principles, which manifest the ethos of the Orthodox Church. There is, first, *the eucharistic ethos*, which, above all else, means using the earth's natural resources with thankfulness, offering them back to God; indeed, not only them, but also ourselves. In the Eucharist, we return to God what is His own: namely, the bread and the wine. Representing the fruits of creation, the bread and wine are no longer imprisoned by a fallen world, but returned as liberated, purified from their fallen state, and capable of receiving the divine presence within themselves. At the same time, we pray for ourselves to be sanctified, because through sin we have fallen away and have betrayed our baptismal promise.

Second, we have *the ascetic ethos* of Orthodoxy that involves fasting and other spiritual works. These make us recognize that everything we take for granted in fact comprise God's gifts provided in order to satisfy our needs. They are not ours to abuse and waste simply because we have the ability to pay for them.

Third, *the liturgical ethos* emphasizes community concern and sharing. We stand before God together; and we hold in common the earthly blessings that He has given to all creatures. Not to share our own wealth with the poor is theft from the poor and deprivation of their means of life; we do not possess our own wealth but theirs, as one of the holy Fathers of

---

9. See the *Office of Vespers for the Preservation of Creation* in *Orthodoxy and Ecology: Resource Book* (Bialystok: Syndesmos, 1996).

the Church reminds us. We stand before the Creator as the Church of God, which, according to Orthodox theology, is the continued incarnate presence of the Lord Jesus Christ on earth; His presence looks to the salvation of the world, not just of humanity but also of the entire creation. The ethos of the Church in all its expressions denotes a reverence for all matter: for the world around us, for other creatures, and for our own bodies.

Hence, our Patriarchal message for this day of protection for the environment is simply that we maintain a consistent attitude of respect in all our attitudes and actions toward the world. We cannot expect to leave no trace on the environment. However, we must choose either to make it reflect our greed and ugliness or else to use it in such a way that its beauty manifests God's handiwork through ours.

September 1, 1995
King, Priest, and Prophet

With the grace of God, we are called once again to celebrate the beginning of the new ecclesiastical year by making special reference, as was established several years ago, to our responsibility toward creation as it relates to the environment within and around us in general and more specifically toward what is referred to as the inanimate natural environment.

Within the context of the continuous and ever developing ecological concern over the years, the holy Great Church of Christ, the Mother Church, was blessed again this past year of 1995, to convene an international seminar on the environment, which was held on the fragrantly scented island of Halki, in the venerable theological school and monastery of the Holy Trinity. This year's seminar had as its main theme "Environment and Ethics." We took advantage of the opportunity to offer certain appropriate introductory remarks—from the perspective of the Mother Church—regarding the sanctity of creation and the lofty responsibility of the human being created in God's image within the whole scheme of creation.

We take this occasion today as well to remind you of what in the seminar we referred to as a temporal sequence in the production of the various species of creation. This temporal sequence etches in stone our responsibility as administrators of God's creation in the world.

The Fathers of the Church, in a manner fitting to God and in appreciation of the temporal sequence mentioned above, taught that every species was created before humankind in order for humankind to enter into a full kingdom and serve there as king, priest, and prophet. We see, therefore, that what might be referred to as the three-fold office of the God-man, about which theology speaks at great length, has been extended to humankind from the very outset and by definition from our primeval relationships with natural creation.

What does it mean for us to reign, minister, and teach in the vast expanse of creation? It means that we must constantly study, serve, and pray to transform what is corruptible into what is incorruptible, to the extent that this can be accomplished during our lifetime.

The Church of the Incarnate Word of God continues His redemptive work in a world that is confused and constantly in a state of ambivalence. Therefore, the Church will never cease to remind the world of these fundamental truths regarding the position and orientation of the human person; rather, by word and deed, the Church teaches a way to life.

We, too, from the holy Great Church of Christ, are striving to fulfill this mission with all our strength. In the context of the celebrations of the 1900th anniversary since the recording of the Apocalypse by St. John on the island of Patmos, we have decided, along with our fellow bishops, to convoke an international ecological symposium. This symposium will sail from Istanbul to the sanctified land of Patmos—where civilizations from the east, west, north, and south have intersected for centuries. Through this means, we shall attempt to reveal more extensively to the eyes of the modern world the magnitude of creation in general, as this was envisioned through God's inspiration by the evangelist of love, St. John the Theologian.

September 1, 1996
A Spiritual, Not Scientific, Problem

Praise and thanksgiving and glory unto our God, venerated in the Holy Trinity, who has deemed us worthy to celebrate once more the commencement of the ecclesiastical year and the traditional feast of the new Indiction.

At the initiative of our Ecumenical Throne, September 1st has been established, as is well known, as a day of prayer and supplication for the protection of our surrounding natural world. Indeed, this has now become established throughout the entire Orthodox world with the consent and accord of the other Most Holy Sister Orthodox Churches. We are hopeful that, with time, the rest of Christendom will come to embrace this proposition and request in order that, before the imminently expiring second millennium in Christ, all Christians may consecrate congruent prayers on the same day. In this way, even from now, in every land inhabited by humankind, we may glorify with thanksgiving in the future the all-holy name of our God, Creator of heaven and earth.

It has become an established tradition, as well, that on this auspicious day of celebration, our Modesty directs anew to the entire congregation of the Church a paternal admonition toward perpetual prayer for our natural environment, stressing in this way all issues deemed timely and necessary.

It is well known to us all that, unfortunately, many such issues arise each year. Naturally, we do not at all overlook the positive efforts made by others on this account. Yet, we observe that the ecological problem has become in many respects more complicated and that the ecological darkness has become even more extensive, which is to say that there is still a substantial ignorance by many and a skillful propaganda on the part of the few, who delight in their alliance with the forces of darkness. These facts result in many fallacies regarding ecology, in purposeful concealment and even distortion of the truth on ecological matters and, indeed, ecological terrorism in the form of exaggeration or abusive intervention in the natural order of things, at times even to the point of exercising interstate threat and violence. This has resulted also in the brutal contravention of international conventions on necessary ecological arrangements and the stubborn refusal to accept the financial burdens of elementary and essential ecological discipline as well as a plethora of other violations, which threaten directly the very air that we breathe.

THE WAY OF RECONCILIATION

All of this may be summarized in the sorrowful realization that, despite the painful current experiences and the concerted efforts of many, extremely few positive steps have been made on the arduous path toward a true and stable reconciliation of humankind with our surrounding physical world.

This failure is due, mainly, to the insistence of the greater part of humankind on the false understanding that the ecological problem is foremost a matter of logical connections, expressed and materialized through the means and methods of politics, economy, technology, and all other human activities.

For all these reasons, it is necessary that the Church steadily call to mind the evangelical truth on this matter, namely that all of the above deviations represent a violation of the divine disposition of the physical world, which cannot remain unpunished given the deviations of such an anti-life stance. It is, indeed, necessary for the Church to remind us that, on the contrary, the aforementioned and therefore imperatively necessary reconciliation, whenever and wherever it is accomplished, represents *par excellence* a spiritual event. More precisely, the Church reminds us that it is the blessed fruit of the Holy Spirit, granted to all who freely and consciously partake in the great mystery of divine love, which has followed the creation, and which constitutes the reconciliation of God, through Christ, with humankind and the entirety of creation.

This reconciliation is understood and experienced within the Church as a settlement: "God in Christ reconciles the world with Himself," not considering their trespasses, as proclaimed by St. Paul, the Apostle to the Nations (2 Cor. 5.19). This is because all things are derived from God, and, in this case, the renewal of all things, "the new creation," in which we become participants through Holy Baptism, as well as other sanctifying gifts in the life of the Church. They are derived from God, who "reconciled us with Himself through Jesus Christ" (2 Cor. 5.18). Moreover, we are all dutifully reminded by the Church that God Himself placed "in us the word of reconciliation" (2 Cor. 5.19), that He entrusted to us Christians the proclamation of the evangelization Gospel of this joyful message of reconciliation, of the new, loving communion of God with humankind and the natural world.

Clearly, this is what is most necessary today. This is what is most urgent for the relationship of humankind with the material world: namely, reconciliation in the aforementioned sense. This reconciliation is not merely to take place in a rational manner, for our benefit and material gain, or, again, for material prosperity and materialistic welfare. It is to be understood theologically, that is with humility and repentance, which lead to true participation in the beauty of creation, restored through Christ, with which the goodness of life is connected harmoniously. And this is so because both goodness and beauty share the same source and cause, which is God Himself, who constitutes the brilliant beauty of the most extreme divine goodness and comeliness.

Indeed, beloved brethren and cherished children in the Lord, this is what is truly good: to participate in that divine goodness and beauty so that we, too, may say, according to Dionysius the Areopagite:

> We say it is good to participate in beauty; indeed, beauty is participation in that cause which beautifies all good things. This supersubstantial good is called beauty due to the loveliness which it transmits to all beings, to each one appropriately.[10]

It is from this divine beauty that the commandment derives that we preserve both the goodness and beauty within us and around us, as the most exalted gifts of God.

For all these reasons, from this sacred center of Orthodoxy, we salute as a significant fact the subject chosen for the second Pan-European Ecumenical Assembly, to be held, God willing, in Austria in June of the coming year of our Savior, 1997: "Reconciliation—gift of God and source of new life." The well-grounded expectation that Christians of Europe will dedicate their efforts to the examination of this issue, both separately and in common, forebodes many blessings. Among these, we expect a clearer sight of divine beauty in nature and a more decisive involvement of all in its protection and further promotion, through the divine gifts and through the multiple creative forces of humankind.

September 1, 1997
Creation and Fall

The all-merciful and beneficent God, who created humanity from non-being into being, so desired that His beloved creature, the king of creation, namely humanity, may enjoy His "very good" creation. Thus God

---

10. *On Divine Names*, PG 3.701. Dionysius was a fifth-century Syriac theologian.

wanted humanity to enjoy creation from the majestic harmony of the heavenly firmament to the natural beauty of the earthly and marine animals as well as all vegetable life, with every harmonious and numerous variation revealed on the surface of the world. For, by contemplating the balance, harmony, and beauty of creation, humanity is lifted to a sense of wonder at the supremely perfect perfection of the divine Creator and, consequently, to love and worship Him. In this way, humanity is sanctified and rendered a partaker of divine blessedness, for which it was destined.

However, the enjoyment of earthly and heavenly things was not granted to people without any presuppositions. Such enjoyment had to be the result also of their own voluntary and active will. In this respect, the first-created[11] were given the command to exercise ascetic abstinence in paradise; they were not to eat of a particular fruit. They were also given the command to labor; they were to work and keep the garden wherein they were placed. Moreover, they were given the command to increase and exercise dominion on the earth, in the same sense in which they were to work and keep the earth as an extended earthly paradise.

These commandments surely also apply to us as the successors of the first-created. They aim, neither solely nor predominantly, at protecting nature in itself, but in preserving the space within which humanity dwells; creation was made for humanity, and it was made beautiful and productive, serving and supporting every goal in accordance to our divine destiny.

Unfortunately, however, the fall of the first-created and the deviation from their goal also resulted in the transformation of their attitude toward nature and their fellow human beings. Thus, today, we are faced with an extremely self-sufficient and greedy behavior of people in relation to the natural environment. Such conduct betrays their indifference toward natural beauty and natural biotopes as well as toward conditions of survival for their fellow human beings.

In order, therefore, to sensitize people in regard to their obligation to contribute as much as they possibly can to the preservation of the natural environment, even if for their own sake, we dedicated September 1st as a day of prayer for the created order. Of course, the natural environment also has value in and of itself. Nevertheless, it acquires greater value when

---

11. "First-created" (or *protoplastoi*) is the conventional manner of referring to Adam and Eve.

contemplated in connection with humanity that dwells within it. For, then, damage to the environment bears consequences not only for nature but also for humanity, rendering human life less tolerable and beautiful.

According to the Church Fathers, a merciful heart will not only seek the heavenly kingdom and sense that it has no abiding city here on earth, seeking instead the heavenly city; it also cannot tolerate any harm to animals and plants, indeed even to the inanimate elements of nature. Such a person recognizes in nature too a relative value, given by God Himself who created it. A similar spirit should also characterize every Christian. We do not limit our expectations simply to this world; nor do we abandon our pursuit of the heavenly reality, namely the divine kingdom. Instead, we recognize that the way that leads to the heavenly Jerusalem goes through the keeping of the divine commandments during our temporary sojourn in this world. Therefore, we are careful to keep the original commandment to preserve creation from all harm, both for our sake and for the sake of our fellow human beings. In any case, respect for the material and natural creation of God, as well as indirectly for all people who are affected by the environment, reveals a profound sensitivity in human attitudes and conduct that should be characteristic of every Christian.

However, because we can discern that such a spirit does not inspire all people, we beseech the Lord on this day especially, that He might illumine them to avoid ecological turmoil and harm. May the Lord, in His long-suffering, protect us all as well as the environment that has, because of our sins, revolted. May He spare us from natural destructions that arise from forces beyond our control.

Let us pray, then, and ask the Lord for favorable and peaceful seasons, free from earthquakes, floods, fires, storms, as well as every raging wind and natural reaction. May He also spare us from every form of human destruction wrought upon our environment, so that we may live in peace and glorify in thanksgiving our Lord Jesus Christ, who bestows every good thing. May His grace be with you all.

September 1, 1998
Creation and Idol

The Holy Orthodox Church, accepting that the entire creation is very good, finds itself in a harmonious relationship with the natural world, which surrounds the crown of creation, the human being.[12] Even though the human being, either as an isolated individual or as collective humanity, is only a minuscule speck in the face of the immense universe, it is a fact that the entire universe is endowed with meaning by the very presence of humanity within it. Based on this assurance, even leading contemporary scientists accept that the universe is infused with the so-called "human principle," meaning that it came about and exists for the sake of humanity.

Consequently, the stance of humanity before its Creator, the all-good God, should have been one of thanksgiving for the abundant wealth, which our Maker has placed at our disposal. However, humanity loved creation more than the Creator and did not return its debt to God. Rather, humanity made an idol of it and desired to be transformed into a wasteful ruler of creation, without any accountability before it, instead of being a rational and grateful consumer of creation. Moreover, humanity was often not even satisfied with wasteful manipulation but schemed to use the tremendous forces contained within nature for the destruction of its fellow human beings and the depletion even of nature itself. From the earliest of days, when Cain murdered Abel, at which point humanity altered the staff formerly used for support into a rod of assault, humanity now tries to use every element as a weapon. Thus, humanity was not satisfied with using elements which God granted in abundance—such as copper, bronze, and iron, and so on—in order to produce tools for a peaceful life. Rather, using all the latest scientific discoveries, humanity fashioned from these elements weapons of mass murder and a system of human annihilation. Unfortunately, humanity continues to make and use these weapons. We, therefore, see gunpowder, nitroglycerine, atomic and nuclear energy, chemical gases, bacterial and every kind of micro-organism and disease-causing factors, being mobilized and gathered into super-modern arsenals, for the purpose of being used as a threat to coerce others into submission but also as a means of active annihilation of those who do not submit.

---

12. This text also served as the basis for remarks of welcome by Patriarch Bartholomew during a colloquium entitled "Monotheism and Environment" organized by the Institute of Religious History and Law of Aix-Marseille University and hosted by the Ecumenical Patriarchate in Istanbul (September 5–8, 1998).

Consequently, neither is the rebellion of nature against humanity a strange coincidence, nor is the continuous exhortation of the Orthodox Church that we should not love the world, which has been led astray from its divine purpose and those things in the world, but that we should love God instead (1 Jn 2.15). In this way, we are able to enjoy the things of the world with blessing and thanksgiving in Christ, through whom we have received reconciliation (Rom. 5.11).

THE REBELLION OF NATURE

Nature rebelled against humanity, which abuses it. Therefore, it no longer finds itself in that perfect divine harmony, whose marvelous melody comes from the rhythmic orbits of the heavenly bodies and the changing seasons of the year. Were it not for the good souls of the saints, who hold together the cohesion of the world, perhaps the revenge of nature for the inhumanities we force it to bear upon our fellow human beings, would be even more lamentable for those people who improperly use its powers against others.

In light of the above, on this first day of September in the year 1998 of our salvation, once again dedicated to the natural environment, we invite and urge everyone to convert the tremendous destructive forces, which we have accumulated on earth—a planet so small in size, yet so great in evil as well as in insurmountable virtue—into creative and peaceful forces.

Unfortunately, the coercion of nature to act destructively against itself and the human race does not come out of the will of certain evil leaders, as much as it is supported by those who wish to deny their own responsibility. It also comes from the consenting will of thousands of individuals, without whose psychological support these leaders would not be able to accomplish anything. Consequently, the responsibility of every living person on the face of the earth flows out of the conscious acceptance or rejection of what has been accomplished. It is through this acceptance or rejection that one participates in the formation of the predominant will. From this point of view, everyone, even the feeblest, can contribute to the restoration of the harmonious renewed operation of the world. We can do so by being in tune with the forces of the divine harmony and not with those which are badly dissonant and oppose the divine all-harmonious rhythm of the universal instrument, of which each one of us constitutes but one of its practically innumerable chords.

Our love for nature does not seek to idolize it; rather, our love for nature stems from our love for the Creator who grants it to us. This love is expressed through offering in thanksgiving all things to God, through whom we, having been reconciled through Jesus Christ (2 Cor. 5.19), enjoy also our reconciliation with nature. Without our reconciliation with God, the forces of nature find themselves in opposition to us. We already experience consequences of this and are subjected to these consequences. Therefore, in order to avert the escalation of evil and to correct that which may already have taken place, and in order to suspend the penalty, we are obligated to accept the fact that we need to be responsible and accountable consumers of nature and not arbitrary rulers of it. We must also accept the fact that, in the final analysis, the demand placed on nature to use its powers in order to destroy our fellow human beings, whom we might consider useless, will result in our facing the same consequences.

Finally, for all these things, we fervently pray to the Lord God that He may show forbearance for our transgressions; that He may grant us time for repentance; and that He may shine in our hearts the light of His truth. This we ask in order that on the issue of the environment, and in each of the paths we encounter before us in life, we may advance in concord with His all-wise, all-harmonious order of the entire creation as it was decreed by Him. Otherwise, our discordant journey may lead to our demise.

September 1, 1999
Creation and Creator

When Paul the Apostle to the Nations advised the Thessalonians to "give thanks in all circumstances" (1 Thess. 5.18), he also counseled them "always to rejoice, and pray without ceasing" (1 Thess. 5.16–17), thus demonstrating that thanksgiving as prayer and everlasting joy go together and coexist inseparably. Truly, one who gives thanks experiences the joy that comes from the appreciation of that for which he or she is thankful; thus, from the overabundance of joy they turn toward the giver and provider of all good things received in grateful thanksgiving. Conversely, the person who does not feel the internal need to thank the Creator and Maker of all good things in this very good world, but instead merely receives them ungratefully and selfishly—by being indifferent toward the one who provided these good things and worshipping the impersonal creation rather than the Creator (Rom. 1.25)—does not feel the deep joy of receiving the gifts of God, but only a sullen and inhumane satisfaction. Such a person is given over to irrational desires, to covetousness, and to "robberies arising from injustice" (Is 61.8), all of which are despised by God. As a result, that person will undergo the breaking "of the pride of his power" (Lev. 26.19) and will be deprived of the sublime, pure, and heavenly joy of the one who gives thanks gratefully.

The belief that every creature of God created for communion with human beings is good when it is received with thanksgiving (1 Tim. 4.3–4) leads to respect for creation out of respect for its Creator. However, it does not fashion an idol out of creation itself. A person who loves the Creator of a given work will neither be disrespectful toward it nor maliciously harm it. Yet, at the same time, that person will surely not worship it and disregard the Creator (Rom. 1.21). Rather, by honoring it, one honors its Creator.

The Ecumenical Patriarchate, having ascertained that the natural creation commonly referred to as "our environment," which in recent times has to a great extent been maliciously harmed, has undertaken an effort that strives to sensitize every person—and especially Christians—to the gravity of this problem for humanity and particularly to its ethical and theological dimension. For this reason, the Patriarchate has established September 1st of each year, which is the natural landmark of the yearly cycle, as a day of prayer for the environment. This prayer, however, is not merely a supplication and petition to God for the protection of the natural environment from the impending catastrophe that is being wrought by humankind on creation. It is also a form of thanksgiving for everything that God in His beneficent providence offers through creation to the good and the wicked alike, to the just and the unjust at once.

The saints of the Christian Church, as well as other sensitive souls—illumined by the divine light that enlightens everyone who comes into the world (Jn 1.9), providing that they sincerely and unselfishly desire to receive this light (Jn 1.11–12)—acquired great sensitivity to all evil that comes to any creature of God and, consequently, to every element that makes up our natural environment.

The saints are models for every faithful Christian to imitate, and their sensitive character is the ideal character toward which we all are obliged to strive. However, because not everyone has this same refinement, those who are responsible for the education of others must continuously teach them what must be done. In light of this, we applaud with great satisfaction the proposal of the Committee on the Environment of the World-wide Federation of Organizations of Engineers, which recently met in Thessalonika during the third international exhibition and conference on Technology of the Environment,[13] that a binding "Global Code of Ethics" for the environment be drafted.

THE ROLE OF THE ECUMENICAL PATRIARCHATE

For its part, the Ecumenical Patriarchate, in addition to proclaiming September 1st as the annual day of prayer for the environment, has successfully organized Symposium II: The Black Sea in Crisis, in collaboration with other interested parties. As a continuation of this effort, the Patriarchate also established the Halki Ecological Institute, which was held successfully this year. This Institute aims at preparing capable people in the countries and churches surrounding the Black Sea to strive in their respective regions to rouse their leaders and people concerning the danger of the impending death of the Black Sea and the general threat of irreparable and harmful damage to the environment. For this reason, the Patriarchate is currently preparing a third international ecological symposium, this time on the Danube, which is a significant source of the pollution for Black Sea, and which has also undergone enormous ecological alterations and disasters as a result of the recent dramatic bombings.

In addition to the ecological and environmental disasters wrought by humankind, natural ones have also occurred, such as the recent earthquakes that have struck Turkey. Despite the fact that, oftentimes, the consequences of these natural occurrences are determined by factors for which humans are responsible, the Church fervently beseeches God to show mercy and compassion on human responsibility, and to show His righteousness and goodness both to those who are responsible as well as to those who are not responsible.

The Ecumenical Patriarchate is fully aware that the end of the second Christian millennium has been sealed by sad and exceptionally destructive

---

13. Organized by HELECO, the initiative of the Technical Chamber of Greece. HELECO is under the auspices of the President of the Hellenic Republic and constitutes the Greek government's legislative organization on technical issues.

occurrences, which transpired mainly in Yugoslavia and Turkey, but which also continue to occur in varying degrees in other parts of our planet. This is principally because the internal spiritual environment of the conscience of each person has not become good, nor has it changed for the better by the grace of God, because of the human ego opposing the beneficent influence of this grace. For this reason, along with the invitation to all that they respect the natural environment, each for his or her own benefit because it is a gift from God to all humankind, the Ecumenical Patriarchate appeals to everyone that they amend their feelings also toward their fellow human beings. Only in this way will the eternal, unchangeable, all-compassionate and merciful God be able positively to influence the free will of the human person and avert the disastrous man-made activities that are upsetting the balance of the environment.

We recognize that heaven and earth pass away, but the laws of God are eternal and unchangeable as is God Himself. Yet, we also know that the law of God is found in the authority of humankind to determine, to a great extent, the path that our life and world will take. For this reason, we summon both ourselves and one other to work toward the common good in all areas, and especially in the area of the environment, which in the final analysis is that realm, which refers first of all to our fellow human beings and subsequently to the natural creation.

September 1, 2001
Harmony Between Matter and Spirit

The designation of September 1st, which marks the beginning of the new ecclesiastical year and of a new period in the physical cycle of social activities, as a day of prayer for the environment, by the Great Church of Christ, the Mother Church of Constantinople, reveals the great significance for humanity of the physical world created by God.

The double nature of the existence of the human person, consisting of body and soul, or of matter and spirit, according to the "very good" and creative will of God, requires the cooperation of humankind and nature, of person and environment. Without this coordination, neither the environment is able to serve humanity according to its destiny nor is humanity able to avert the disturbance of natural balances and the obliteration of the natural harmony, which God created for us.

Unfortunately, because of human desire to gain power and wealth, humanity often trespasses the limits of the endurance of nature, subjecting it to maltreatment or abuse. On other occasions, humanity transgresses again the commandment of God to the first-created to labor and to keep the natural creation, becoming indifferent to the maintenance of its integrity and natural balance.

The result of this behavior is the disturbance of the natural harmony, and the rebellion, as it were, of the impersonal nature, which produces phenomena that are the exact opposite to those that serve humankind's normal life. The radiation of a power that is able to benefit humanity becomes an explosive potency of inconceivable destruction. The rivers that are meant to be bearers of life-giving water become carriers of destructive floods. The explosive potency of dynamite is transformed from a useful instrument to a power of homicide and total ruin. Rain is changed from a means of irrigation of plants and the watering of animals to a cause of drowning and uncontrollable currents. Combustion from energy sources and heat becomes a source of atmospheric pollution. In general, the totality of natural possibilities, designed to serve in regular operation humanity's natural survival, are stretched by humankind beyond their regular limits and as a result awaken the avenging powers of abuse, which are released when the permissible use is transformed into a means of satisfaction of human audacity with a view to the limits of nature.

THE AUDACITY OF BABEL

The audacity of those who built the Tower of Babel produced a break in human understanding and communication.[14] Humanity's exclusive turning to the carnal aspect of being, to the exclusion of the spiritual aspect, brought about the purging cataclysm of Noah's times. Since then, God refrains from letting natural disaster bring humanity back to its senses, as the rainbow symbolizes. Nevertheless, humanity continues to pursue its greedy efforts toward forcing nature to mass production and unnatural usage. As a result, humanity procures terrible environmental disasters that primarily damage

---

14. See Genesis 11.

humanity itself. We may recall here such well-known cases as the environmental calamities incurred by nuclear explosions and radio energy waste, or by toxic rain and polluting spillages. We may also recall the consequences of the violent feeding of vegetarian animals that is enforced by human audacity in order to produce food from animals. All this constitutes an insolent overthrow of natural order. It is indeed becoming generally accepted that the overthrow of this natural order in the personal and social life of human beings produces ill reactions to the human organism, such as the contemporary plagues of humanity, cancer, the syndrome of post-virus fatigue, heart diseases, anxieties and a multitude of other diseases.

All these bear witness to the fact that it is not God but humanity that causes contemporary plagues, which attack our well being, because "man is the most disastrous of disasters," as the ancient tragedy puts it. Thus, if we want to improve the conditions of the material and psychological life of humanity, we are obliged to recognize and to respect the natural order, harmony, and balance, and to avoid causing disarray in the natural powers, which are released when the cohesive bond of the universal and particular harmony, especially of the ecological one, is audaciously overturned. Nature was placed by God to the service of humankind, on the condition, however, that humanity would respect the laws that pertain to it and would work in it and protect it (Gen. 2.15).

On this day, dedicated by the Mother Church to prayer for the natural environment, we supplicate the Lord to restore with His divine and almighty power the natural order, wherever human audacity has overturned it, so that humanity might not suffer the tragic consequences of unlawful violations of nature by human actions. We all share the responsibility for such tragedies because we tolerate those immediately responsible for them and accept a portion of the fruit that results from such an abuse of nature. Consequently, we all need to ask for God's forgiveness and illumination so that we may come to understand the limit that distinguishes the use from the abuse of nature and never trespass it.

September 1, 2002
Unpaid Debt to Nature

When the Mother Church declared September 1st each year as a day of prayer for the environment, no one could imagine at the time (as early as

1990) just how rapidly the natural conditions of the world would deteriorate as a result of harmful human interference or how horrific the consequent losses and damages on human lives would be. Recent floods in Europe, India, and Russia, as well as those occurring during this year or previously in other parts of the earth, all bear witness to the disturbance of the climactic conditions caused by global warming. Such disasters have persuaded even the most incredulous persons that the problem is real, that the cost of repairing damages is comparable to the cost of preventing them, and that there is simply no margin left for remaining silent.

The Orthodox Church is a pioneer in her love and interest for humanity and its living conditions. Therefore, on the one hand, it recommends that we lead virtuous lives, looking to eternal life in the heavenly world beyond. On the other hand, however, it also recognizes that—according to the teaching of our Lord Jesus Christ—our virtue will not be assessed on the basis of individualistic criteria, but on the basis of applied solidarity. This is so characteristically described in the parable of the Last Judgment (see Matthew 25). In this parable, the criterion for being saved and inheriting the eternal Kingdom is supplying food to the hungry, clothes to the naked, aid to the sick, compassion to prisoners. Generally, the criterion is offering our fellow human beings the possibility of living on our planet under normal conditions and of coming to know God in order to enter His Kingdom.

This means that the protection of our fellow human beings from destructive floods, fires, storms, tempests and other such disasters is our binding duty. Consequently, our failure to assume appropriate measures for avoiding such phenomena is reckoned as an unpaid debt and constitutes a crime of negligence, incurring a plethora of other crimes, such as the death of innocent people, the destruction of cultural monuments and property as well as overall regressive progress. We pray, therefore, that God will remove natural destruction, which we cannot avert by our own care and foresight. Yet, at the same time, it is our obligation to engage in the labor of study and the expense of securing necessary measures in order to avoid such disasters as are derived from wrongful human action.

It is true that the great part of such measures and expenditures cannot be taken from isolated individuals because they transcend their capabilities. Sometimes, they even transcend the capabilities of individual states, requiring the cooperation of several states and even of humanity in its entirety. Thus, we heartily salute the international consultations on this matter, which are taking place throughout the world or will take place in the future, and pray that they may conclude their deliberations with unanimous decisions on the measures that should be taken as well as on their implementation.

Nevertheless, what contributes most of all to the creation of this ecological crisis is the excessive waste of energy by isolated individuals. The restriction of wasteful consumption will blunt the acuteness of the problem, while the constant increase in the use of renewable sources of energy will intermittently contribute to its alleviation. However insignificant the contribution of any individual in averting further catastrophe to the natural phenomena, we are all called and obliged as individuals to do whatever we can. Only then can we confidently pray to God that He may supply what is lacking in our effort and efficacy.

We paternally urge everyone to come to the realization of their personal responsibility and to do all that is possible in order to avert global warming and environmental aggravation. We fervently entreat God to look favorably on the common effort of all and to prevent further imminent disasters on our natural environment, within which He ordered us to live and to fight the good fight so that we may inherit the heavenly Kingdom.

September 1, 2003
Extreme Weather and Extremist Behavior

This day of prayer for the environment finds us disturbed by information recently received about two distressing events. The first is the severe destruction of the natural environment owing to rising temperatures, which has caused extensive forest fires in some parts of the world. The second refers to the outbreak of collective killings of citizens that were innocent and unconnected with their killers, executed by extremist elements whose aim is to harm the society to which these randomly selected victims belong.

In both cases it is obvious that humankind is responsible, as is the perversity that causes it. On one hand, these climatic irregularities and their resultant natural catastrophes are mainly caused by human activities, which are performed without thinking about or making any prior assessment of the devastating effects they have on nature. On the other hand, the collective execution of innocent and arbitrarily chosen people with the aim of spreading fear throughout society and forcing it to give in to the just or unjust demands of these executioners implies an ignorance of the fact that this method has been used many times in the past but has never succeeded in its purpose. Therefore, if a rational assessment of the results of these actions had prevailed amongst the perpetrators, then a lot of suffering would have been avoided. Many desperate actions against innocents could also have been avoided if the wrong doers recognized and respected the rights of the wronged.

The natural environment was created by God to be friendly and of service to the needs of humankind. However, owing to Adam's original disobedience, the natural harmony and balance of the environment was disrupted and because of his persistent disobeying of God's commandments, it continues to disrupt, leading to total disarray and disharmony. Therefore, the prayer that we offer up to the Lord today for the protection of the natural environment from all kinds of destruction and disruption, should first of all be a prayer for the repentance of man, who through misjudged, thoughtless, and sometimes arrogant actions directly or indirectly provokes most, not to say all, natural catastrophes.

Our Lord who taught us the Lord's Prayer includes in it a promise that accompanies a request: "Forgive us our trespasses as we forgive those who trespass against us." This has a broader meaning. Our prayer should be accompanied by a corresponding sacrifice, mainly a sacrifice of our selfishness and arrogant pursuits, which demonstrate our insolent attitude toward the Creator and His wisely stipulated natural and spiritual laws. This change of attitude and mentality is called repentance. Only if our prayer to God for the protection of the environment is accompanied by correspondent repentance will it be effective and welcomed by God.

Therefore, beloved brothers and sisters in the Lord, let us reconsider our lives and let us repent for everything we do mistakenly and against the wise laws of God, in order to be heard by Him, begging His kindness to maintain the natural environment, friendly and undamaged for humankind. May the Lord God grant stability to preserve the natural world, which He created full of grace and harmony for the sake of humanity.

September 1, 2004
What Have We Achieved?

Fifteen years ago, our venerable predecessor, the late Patriarch Dimitrios issued the first official decree for the preservation of the natural environment, an encyclical letter to the *pleroma* of the Church, formally establishing September 1st as a day of prayer for the protection of the environment.[15] That historical proclamation emphasized the significance of the eucharistic and the ascetic ethos of our tradition, which provide a corrective for a consumer lifestyle and an alternative to the prevailing philosophy of our age.

The Church Fathers have always insisted on the critical importance of self-examination as a pre-condition for spiritual growth. Echoing the classical oracle of Delphi, Clement of Alexandria[16] exhorts: "Know yourself! If you know yourself, you will know all things." Evagrius of Pontus[17] states: "He who knows himself knows God." And Isaac the Syrian[18] claims: "To know oneself is to know one's failures, which leads to the resurrection of the dead." Therefore, let us consider what we have learned as a Church over the last fifteen years. What knowledge have we gained? What failures have we experienced? And what direction should we now assume?

In the five summer seminars that were held annually from 1994 to 1998 at the Theological School on the Island of Halki, we learned about the close connections between environmental issues and education, ethics, communication, justice, and poverty. And in the five international symposia held biennially from 1995 to date, we have explored the impact of our wasteful lifestyle on the waters of the Aegean Sea, the Black Sea, the Danube River, the Adriatic Sea, and the Baltic Sea. Together with theologians, scientists, politicians and journalists, we recognized in a tangible manner the responsibility that we all bear—before one another, before our world, and before our God—for the destruction of our world's natural beauty, for the depletion of the earth's resources, and for the devastation of our planet's diversity.

---

15. See the first encyclical message in this chapter.
16. Clement (150–215) was head of the famous School of Alexandria.
17. Evagrius (345–399) was an ascetic and monastic author.
18. Isaac of Nineveh was a seventh-century mystic and author.

More especially, we have appreciated how the preservation of the natural environment is intimately related to the cessation of warfare, to the restoration of social justice and to the management of world poverty. We have learned how the way that we treat human beings is directly reflected in the way that we relate to the natural environment, as well as to the worship that we exclusively reserve for God. It should come as no surprise to us that we are able to misuse the natural and material creation when we are able to abuse our fellow-human beings. The Mother Church has been at the forefront of significant gatherings and agreements of world peace and welfare, of economic and social reform, of human rights and religious tolerance.

STRENGTH IN THEORY, WEAKNESS IN APPLICATION

When it comes to the appropriate response and the proper theological reflection, there is no doubt that our Orthodox Church has a great deal to contribute to the contemporary debate concerning ecology. We are able to draw upon the depth and wealth of our Scriptural and Patristic heritage in order to contribute positively and constructively to the critical issues of our time. Where, however, as Orthodox Christians we reveal the greatest vulnerability is in the practice of our theory.

It is always the easier approach to lay blame on Western development and technological progress for the ills that we confront in our world. And it is always a temptation to believe that we hold the solution to problems that we all face today or else to ignore the imminent danger that we face globally. What is more difficult—and yet at the same time more noble—is to discern the degree to which we constitute part of the problem itself. Just how many of us examine the foods that we consume, the goods that we purchase, the energy that we waste, or the consequences of our privileged living? How often do we take the time to scrutinize the choices that we make on a daily basis, whether as individuals, as institutions, as parishes, as communities, as societies, and even as nations?

More importantly, just how many of our Orthodox clergy are prepared to assume leadership on issues concerning the environment? How many of our Orthodox parishes and communities are prepared to materialize the knowledge that we have accumulated in recent years by practicing ecologically sensitive principles in their own communities? How do the decisions of any local community and parish reflect on a practical and tangible level the experience that we have gained on a theoretical level?

In an age when the information is readily available to us, there is surely no excuse for ignorance or indifference. To overlook is to shut our eyes to a reality that is ever-present and ever-increasing. Former generations and cultures may have been unaware of the implications of their actions. Nevertheless, today, more perhaps than any other time or age, we are in a unique position. Today, we stand at a crossroads, namely at a point of choosing the cross that we have to bear. For, today, we know fully well the ecological and global impact of our decisions and actions, irrespective of how minimal or insignificant these may be.

It is our sincere hope and fervent prayer that in the years ahead, more and more of our Orthodox faithful will recognize the importance of a crusade for our environment, which we have so selfishly ignored. This vision, we are convinced, will only benefit future generations by leaving behind a cleaner, better world. We owe it to our Creator. And we owe it to our children.

September 1, 2005
Doing, Not Just Saying

> The earth, having no tongue cries out sighing, why are you people polluting me with many evil things? (*9th Ode, Earthquake of October 26*)

In a very pictorial way, the holy hymnographer Joseph[19] presents the earth as grieving and protesting voicelessly for the many evils with which we burden her. If this holy hymnographer thought back then that the pollution of earth by humankind would cause the wrath of God, today, humanity in its entirety should all the more realize our ultimate destructive behavior against the creation of God.

Certainly, the earth was created well-equipped to offer shelter to the human beings and was ordered by God to cover their needs. However, we do not draw from earth's resources what we need in moderation, so that we allow its productive ability to remain sound and intact; instead, we are depleting her natural resources. We draw so much to such excess

---

19. A ninth-century monk and liturgical poet of Byzantium.

and in such rough ways that we weaken her abilities and destroy all future production of natural resources. In doing so, we resemble those who act greedily, and who, when in need of collecting wood, destroy both the trees and the forest and, thus, deprive themselves of the opportunity to collect more wood in the future. It is a known historical fact that many areas of the earth that had once been sites of developed civilizations, ended up in total devastation.

This phenomenon of devastation, which unfolded slowly in earlier times, is progressing in our times at a high speed. Vast expanses around polluting factories and industrial zones that emit toxic waste have already been deadened, and the number of such dead expanses is constantly growing. Huge regions have been made subject to deforestation in order to be used as cultivation grounds, but the utilization of toxic pesticides has destroyed any form of sprouting, except for the object needed. These non-biodegradable toxic pesticides enter the water-air cycle and pollute the springs and rivers causing severe problems to human health. In regard to the consequences of these methods employed and materials used, greed and negligence take their revenge. While we work hard to increase the productive ability of our planet, we, on the other hand, destroy it. Astronauts, who have observed the whole earth from a distance while in orbit, drew humankind's attention to the fact that huge expanses of it have been deforested and will end up in devastation.

The aforementioned holy hymnographer Joseph personifies earth, which, addressing man, complains that the Master of humankind and God whips her instead of him, for God wants to spare the human being; the earth, however, bemoans her suffering due to humankind's mistakes and cries to people: "Come to your senses and appease God in repentance." This invitation is quite timely. We must realize the forthcoming danger; we must understand its causes and acknowledge our responsibility. We must aim to appease God, not through words and small sacrifices, but through courageous acts and large sacrifices. For the promise of the Lord that we will receive back in multiple that which we sacrifice, applies here as well.

The Mother Church is fully aware of the dangers that threaten the earth and our surrounding natural environment brought about by both the natural ramifications of human acts but, mostly, by the moral consequences of human crimes; therefore, the Mother Church established the

1st of September as a day of prayer for the environment. Prayer appeases God; however there is also validity in the saying of ancient Greeks: "In addition to asking for help from goddess Athena offer also your own effort." This saying is similar to the biblical phrase "The effectual fervent prayer of a righteous man avails much" (James 5.16). The importance of this sentence lies on the word "effectual," which means that the prayer is more powerful when accompanied by actions for the one for whom we are praying. For there is no vindication for the one who says "Lord, Lord," but rather for the one who does the will of God. In our discussion, it is evident that the will of God mandates the preservation of the eternal yielding of our natural resources, respect toward the natural creation of God and our future generations, and the reversal of our destructive behavior against the very good natural environment that was given to us by God.

September 1, 2006
Catastrophic Arrogance

The God of tender mercy and love for mankind created the cosmos to be a place of sublime beauty, serviceable and apt to the needs of every human being. Into such a world, God allowed the crown and monarch of His creation, the human person, to partake of everything in it that is needful for life.

Every necessary relationship of humanity with creation is conjoined with a sense of joy and satisfaction. If there is an excess or privation of what is, by its use, naturally good then there is an accompanying sensation of want (in the case of privation) or surfeit (in the case of excess). Thus does the human being possess in himself instinctively a means of measuring beneficial need or detrimental excess. The need manifests as privation; the excess manifests as wasteful superfluity. It follows then that all human beings, endowed with freedom of will, have the capacity to direct their own instinctual faculty to prescribe their own limits; whether to restrain such limits for reasons of ascetical discipline, or to exceed them by the power of desire.

Thus we find ourselves confronting this condition: either we are subject to greed (which is idolatry according to the Apostle Paul [Col. 3.5]), or to a certain hatred for life, for the natural blessings and gifts God, that is to attitudes which are equally unacceptable, being opposite to the perfect plan of God for humankind's enjoyment of life.

The unfortunate reality is that humanity has rejected to be shaped by the suggestions and inducements of God. We have not followed His guiding grace in determining the measure of our needs and how we use the world; how we work in the world or how we preserve the world. The result is that we behave toward the environment, toward nature, rapaciously and catastrophically. When we apply our own sense of mastery and not appropriate use we upset the natural harmony and equilibrium that is based in God. Nature reacts negatively and the result is that terrible desires pile up on the human family. Recent unusual fluctuations in temperature, typhoons, earthquakes, violent storms, the pollution of the seas and rivers, and the many other catastrophic actions for man and the environment ought to be an obvious alarm for something to be done with human behavior. The principal reason for this catastrophic behavior of contemporary man is his egocentrism, which is another face of self-reliance apart from God, and even self-divinization.

On account of this egocentrism, the relationship between humanity and nature has been radically altered. Now an impertinent, arrogant subjugation of the forces of nature has supplanted that which was designed by God. In place of the preservation of life and freedom, these forces serve to destroy and oppress our fellow man, or we indulge in excessive consumption, without regard to the consequences of such excess.

The use of atomic and nuclear forces of nature for warlike purposes constitutes unmitigated hubris. Whatever the manner of our over consumption, we have burdened the natural environment with such pollution that the earth's temperature is rising and many of nature's balancing acts are now unstable, with all that this implies. The enormous amount of energy that is consumed for the purposes of the modem war-machine, as well as the prodigality of modem life that far exceeds the reasonable human needs of today, comprise two distinct sectors, in which the responsibilities of leaders and simple citizen are woven together in such a way that each has the capability of taking action for the betterment of the general condition.

Beloved children and brethren in the Lord, let us take action, each one from his own position and setting, giving every effort to an amelioration of senseless consumption. Let us work toward a restoration of a harmonious working of the planet on which we live, so that in tranquility our children may enjoy all the blessings of the creation of our loving God, the blessing He offers to all people.

September 1, 2008
Creation Continues to Groan

> For creation was made subject to vanity, not willingly, but by reason of him who subjected it . . . For we know that all of creation groans and travails in pain together until now. (Rom. 8.20, 22)

Once again, as the ecclesiastical year begins, we are called to reflect—with renewed spiritual intensity in Christ and especial sensitivity—on the state of our bountiful planet, and to offer particular prayers for the protection of the whole natural world.

Many things have changed since our predecessor, the late Patriarch Dimitrios decided, over two decades ago, that September 1st should be dedicated as a day of supplication for the preservation of God's beautiful creation. In assuming that initiative, the late Ecumenical Patriarch also issued a message of warning about the destructive consequences of abusing the environment. He noted that, in contrast with most other forms of human misuse and violation, environmental pollution has the potential to cause vast and irreversible damage, by destroying virtually all forms of life on the planet.

At the time, of course, this warning may have sounded exaggerated to certain skeptical ears; however, in the light of what we know now, it is abundantly clear that his words were prophetic. Today, environmental scientists expressly emphasize that the observed climate change has the potential to disrupt and destroy the entire ecosystem, which sustains not only the human species but also the entire wondrous world of animals and plants that is interdependent upon one another like a chain. The choices and actions of what is otherwise civilized modern man have led to this tragic situation, essentially comprising a moral and spiritual problem, which the divinely inspired Apostle Paul had articulated with colorful imagery in underlining its specifically ontological dimension in his Letter to the Romans nineteen centuries ago: "Creation was made subject to vanity, not willingly, but by reason of him who subjected it . . . For we know that all of creation groans and travails in pain together until now" (Rom. 8.20, 22).

AWARENESS IS INCREASING, ACTION IS DECREASING

At this point, however, we are obligated to state that this spiritual and moral dimension of the environmental problem constitutes today, perhaps more so than ever before, the common conscience of all people, and especially young people, who are well aware of the fact that all of humanity has a common destiny. An increasing number of people comprehend that their overall consumption—namely, their personal involvement in the production of particular goods or their rejection of others—touches not only on ethical, but also on eschatological parameters. An increasing number of people understand that the irrational use of natural resources and the unchecked consumption of energy contribute to the reality of climate change, with consequences on the life and survival of humanity created in the image of God and is therefore tantamount to sin. An increasing number of people characterize either virtuous or else vicious those who correspondingly treat created nature either reasonably or unreasonably.

Nevertheless, by the same token, even as people's awareness of the environmental crisis grows, unfortunately the image presented by our planet today is the opposite. Especially disturbing is the fact that the poorest and most vulnerable members of the human race are being affected by environmental problems which they did not create. From Australia to the Cape Horn of Africa, we learn of regions experiencing prolonged drought, which results in the desertification of formerly fertile and productive areas, where the local populations suffer from extreme hunger and thirst. From Latin America to the heart of Eurasia, we hear of melting glaciers, on which millions of people depend for water supply.

Our Holy and Great Church of Christ, following in the footsteps and example of the late Ecumenical Patriarch Dimitrios, is working tirelessly to raise awareness not only among public opinion but also among responsible world leaders. It achieves this by organizing ecological symposia that deal with climate change and the management of water. The ultimate purpose of this endeavor is to explore the interconnectedness of the world's ecosystems and to study the way in which the phenomena of global warming and its anthropogenic effects are manifested. Through these academic gatherings, attended by representatives of various Christian churches and world religions as well as diverse scholarly disciplines, our Ecumenical Patriarchate is striving to establish a stable and innovative alliance between religion and science, based on the fundamental principle that—in order for the goal to be achieved and for the natural environment to be preserved—both sides must show a spirit of good will, mutual respect, and cooperation. The collaboration of science and religion at these symposia organized in different regions of the planet, seeks to contribute to the development of an environmental ethic, which must underline that the use of the world and the enjoyment of material goods must be eucharistic, accompanied by doxology toward God; by the same token, the abuse of the world and participation therein without reference to God is sinful both before the Creator and before humanity as creation.

Beloved brothers and children in the Lord, we know that the creation participated in the fall of Adam from the original beauty; as a result, it groans and travails in pain together. Moreover, we know that the abuse, deviation, violation, and arrogance of humanity contribute to the destruction of the travailing nature, which is subjected to the corruption of creation. Finally, we also know that this destruction actually comprises self-destruction. Therefore, we invite all of you, irrespective of position and profession, to remain faithful to a natural use of all God's creation, "offering thanks to the God, who created the world and granted everything to us." For to Him is due all glory and power to the ages.

September 1, 2009
The Global Market and the Natural Environment

As we come again to the changing of the Church year, we reflect once more on the state of God's creation. We think about the past and repent for all that we have done or failed to do for the earth's care; we look to the future and pray for wisdom to guide us in all that we think or do.

These last twelve months have been a time of great uncertainty for the whole world. The financial systems that so many people trusted to bring them the good things of life, have brought instead fear, uncertainty, and poverty. Our globalized economy has meant that everyone—even the poorest who are far removed from the dealings of big business—has been affected.

The present crisis offers an opportunity for us to deal with the problems in a different way because the methods that created these problems cannot provide their best solution. We need to bring love into all our dealings, the love that inspires courage and compassion. Human progress is not just the accumulation of wealth and the thoughtless consumption of the earth's resources. The way that the present crisis has been dealt with has revealed the values of the few who are shaping the destiny of our society; of those who can find vast sums of money to support the financial system that has betrayed them, but are not willing to allot even the least portion of that money to remedy the piteous state the creation has been reduced to because of these very values, or for feeding the hungry of the world, or for securing safe drinking water for the thirsty, who are also victims of those values. On the face of every hungry child is written a question for us, and we must not turn away to avoid the answer. Why has this happened? Is it a problem of human inability or of human will?

We have rendered the global market the center of our interest, our activities and, finally, of our life, forgetting that this choice of ours will affect the lives of future generations, limiting the number of their choices that would probably be more oriented toward the well-being of man as well as the creation. Our human economy, which has made us consumers, is failing. The divine economy, which has made us in the image of the loving Creator, calls us to love and care for all creation. The image we have of ourselves is reflected in the way we treat the creation. If we believe that we are no more than consumers, then we shall seek fulfillment in consuming the whole earth; but if we believe we are made in the image of God, we shall act with care and compassion, striving to become what we are created to be.

Let us pray for God's blessing on the United Nations Climate Change Conference in Copenhagen in December,[20] so that the industrially developed countries may cooperate with developing countries in reducing harmful polluting emissions, that there may exist the will to raise and manage wisely the funds required for the necessary measures, and that all may work together to ensure that our children enjoy the goods of the

---

20. See Chapter 9 of this book, "Declarations and Statements."

earth that we leave behind for them. There must be justice and love in all aspects of economic activity; profit—and especially short-term profit— cannot and should not be the sole motive of our actions.

Let us all renew our commitment to work together and bring about the changes we pray for, to reject everything that is harming the creation, to alter the way we think and thus drastically to alter the way we live.

September 1, 2010
The Financial Crisis and the Ecological Crisis

Our ever-memorable predecessor, the late Patriarch Dimitrios, who possessed a deep awareness of the gravity of the environmental crisis, as well as of the responsibility of the Church to directly and effectively confront the crisis, issued the first official encyclical dealing with the protection of the natural environment more than two decades ago.[21] Through this encyclical, the Mother Church officially established the date of September 1st—the beginning of the ecclesiastical year—as a day of prayer for the protection of the environment, declaring it to the plenitude of the Church throughout the length and breadth of the world.

At that time, our Church insightfully emphasized the significance of the eucharistic and ascetic ethos of our tradition, that manifests our most important and most crucial unique contribution toward the proper and universal struggle for the protection of the natural environment as a Divine Creation and shared inheritance. Today, in the midst of an unprecedented financial crisis, humanity is facing many and diverse trials. But this trial is related not only to our individual hardships; this trial affects every aspect of human society, especially our behavior and perception of the surrounding world and the way we rank our values and priorities.

It is important to note that the current grievous financial crisis may spark the much-reported and absolutely essential shift to environmentally viable development, i.e., to a standard of economic and social policy whose priority will be the environment, and not unbridled financial gain. Let us all consider as an example what may happen to countries that are suffering today on account of the financial crisis and poverty, such as Greece, which at the same time have exceptional natural riches: unique

---

21. See the first section of this chapter.

ecosystems, rare fauna and flora and natural resources, exquisite land-scapes, and abundant sunlight and wind. If ecosystems deteriorate and disappear, natural sources become depleted, landscapes suffer destruction, and climate change produces unpredictable weather conditions, on what basis will the financial future of these countries and the planet as a whole depend?

We hold, therefore, that there is a dire need in our day for a com-bination of societal sanctions and political initiatives, such that there is a powerful change in direction, to a path of viable and sustainable environ-mental development.

For our Orthodox Church, the protection of the environment, as a divine and very good creation, embodies a great responsibility for every human person, regardless of material or financial benefits. The direct cor-relation of the God-given duty and mandate, to work and preserve, with every aspect of contemporary life constitutes the only way to a harmoni-ous co-existence with each and every element of creation, and the entirety of the natural world in general.

Therefore, we call upon all of you, beloved brethren and children in the Lord, to take part in the titanic and righteous battle to alleviate the environmental crisis, and to prevent the even worse results that derive from its consequences. Let us motivate ourselves to harmonize our per-sonal and collective life and attitudes with the needs of nature's ecosys-tems, so that every kind of fauna and flora in the world and in the universe may live and thrive and be preserved.

2

Orthodox Theology
and the Environment

General Addresses

AWARD CEREMONIES

*Address at the Conferral of the First Honorary Doctorate of the
Department of Environmental Studies, University of the Aegean,
Mytilene, Greece, October 27, 1994*

ORTHODOXY AND THE ENVIRONMENT

Any effort to connect contemporary environmental pursuits to theological presuppositions seems to be a paradoxical and perhaps eccentric venture. In the minds of most, ecology represents a practical and tangible methodology. On the contrary, theology and theological cosmology, even as terminology, are for most people naturally connected to abstract theoretical pursuits. They refer to associations of doctrines and ideologies that are irrelevant to daily life and its problems.

Contemporary ecology, as a matter of scientific study, but also in the form of crusades and movements for the salvation of our earthly ecosystem, is one of the most characteristic expressions of human interest concentrated on practical goals. The logic of environmental protection is presented as a purely utilitarian matter. If we do not protect our natural environment, then our own survival will increasingly be rendered more difficult and problematic, while the very presence of the human race on this planet will be threatened very soon. The danger of degeneration or even of annihilation of the human race is described as imminent.

THE LOGIC OF ECOLOGY AND THE LOGIC OF THEOLOGY

In the context of this apparent way of thinking, the natural environment is conceived as a necessary and sufficient condition for human existence and survival; yet, this condition or context is perceived as utilitarian. The method of thinking is limited to the manner of usage. What is of interest is not the source or cause of the natural reality; nor is any hermeneutical "meaning" sought in the cosmic order and harmony, or in the wisdom and beauty of nature. It is, of course, quite possible that those things that exist in nature were created by some unknown "higher power," or that they are the products of inexplicable "chance" and automatic force— again without explanation—that exist innately in the very structure of matter. At any rate, the interpretation of any cause or end is not what gives meaning to everything that exists.

Based, then, on such a utilitarian logic, contemporary ecological movements demand the definition of rules for the use of nature by humanity. Ecology seeks to be seen as a practical ethic of human conduct in relation to the environment. Nevertheless, like any other ethic, ecology too provokes us to ask the question: Who is it that defines these rules of human conduct, and with what authority are they defined? What logic renders these rules obligatory? From what source do they derive their validity?

The rationalism of this utilitarian mentality is perhaps the only answer that ecologists can provide to this question. The correctness of an ecological ethic is derived from and based on its apparent utilitarianism. It is reasonably beyond any doubt that, in order for the human race to survive on this earth, there are certain conditions that are necessary also for the natural environment.

Nevertheless, the rationalism of this utilitarian ethic is precisely what led to the destruction of the natural environment. Humanity has not destroyed the environment out of some senseless masochism. Humanity destroys the environment in an effort to exploit nature, in order to secure more facilities and comforts in daily life. The logic that led to the destruction of the environment is precisely the same logic now as that concerning the protection of the environment. Both "systems of logic" approach nature as something exclusively utilitarian. Neither attributes any different meaning to nature. Both exist on the same level of an ontological interpretation of the natural reality, or rather on the vacuum of deliberate ignorance of any ontological interpretation.

Consequently, the difference revealed by these two "systems of logic" (namely, that which leads to the destruction and that which looks to the protection of the environment) is only quantitative. Ecologists demand a limited and controlled use of the natural environment, a quantitative reduction that will allow its longer use by humanity. They seek a rational limitation of our limitless use of nature. Therefore, they seek a more rational application of a rational system that already exploits nature. They seek a utilitarian restraint in our utilitarian abuse.

Yet, who will define this quantitatively lesser and more correct use of nature within the context of the very same logic? By what measures will this be defined? Although the goal appears to be extremely rational, in fact it is by definition as well as in practice irrational. By definition, it is self-contradictory inasmuch as utilitarianism cannot work against utilitarianism. In practice, it is irrational because the majority of this earth's population does not accept to be deprived of facilities and comforts secured for the convenience of a very small minority in "advanced" societies through the destruction of the environment.

In order, then, to meet the demands of ecologists, another logic is required, one that is able to substitute the logic of utilitarianism. Their demand must be established on an entirely different basis. For example, it may be rooted in an altruism that is concerned for the survival of future generations, or else based on the demand for some "quality" of life that is not measured by consumer comfort and greed. Therefore, we need to find a basis that is not utilitarian but is universally accepted. And it is impossible to define non-utilitarian goals, on which everyone would agree, based on purely rational criteria. We must discern within the human person different needs and a different hierarchy of values. This can only happen when the human conscience acknowledges a different meaning in life and the world, one that is not exclusively utilitarian.

A RELIGIOUS LOGIC AND A RELIGIOUS ETHIC

The monotheistic religions preserve an appreciation of the natural reality that is not exclusively utilitarian. According to their traditions, the world is the creation of God. The use, then, of the world by humanity constitutes a practical relationship between humanity and God because God bestows and humanity receives the natural goods as an expression of divine love.

Two fundamental consequences follow from this understanding. First, *the use of the world is not an end in itself for humanity, but a way of relating to God.* If humanity distorts the use of this world into an egocentric abuse of greed, by dominating and destroying nature, then humanity is denying and destroying its own life-giving relationship with God, a relationship destined to continue into eternity.

Second, *the world, as God's creation, ceases to be a neutral object for human use.* It incarnates the word of the Creator like every other creation embodies the word of its artist. The objects of natural reality bear the seal of their divine Creator's wisdom and love; they are words (*logoi*, which also implies meaning) of God inviting humanity to dialogue (*dialogos*) with God.

In spite of all this, it is a given historical fact that the contemporary concept of the world as a neutral object that may be used and exploited by humanity for its own individualistic pleasure, arose and was articulated within the context of Christian Europe. It would take a special analysis of the historical circumstances and theoretical presuppositions that led Christian Europe to replace the relationship between humanity and the world with the understanding that humanity has limitless domination over the world. The reasons for the divergence are not unrelated to the causes that led to the painful schism in the eleventh century, namely to the severance of Western Christianity from the unified body of the One, Holy Catholic and Apostolic Church.

It nevertheless remains a fact that the change in human behavior toward the environment today requires a change in meaning, namely a change in the meaning attributed by human beings to matter and nature. Ecology will not inspire any respect toward nature if it is not informed by a different cosmology to that which prevails today in our culture and civilization, if it is not exempted from a naïve materialism and an equally naïve idealism alike.

AN ORTHODOX COSMOLOGY

Let us endeavor to underline in brief the potential of an Orthodox ecclesiastical cosmology to contribute toward the discovery of a new and different logic. Our general remarks will be primarily based on the works of St.

Gregory of Nyssa,[1] St. Maximus the Confessor,[2] and St. Gregory Palamas.[3] However, this does not imply that other patristic writers do not substantially contribute to this discussion.

The fundamental patristic contribution here is the introduction of a third ontological category for the interpretation of existent reality and its principal source. We are referring to the category of divine energies, which are added to the couplet known as divine essence and divine hypostasis, which prevails in ontological philosophical analysis. Although the starting-point of ecclesiastical thought was primarily theological, we shall draw upon the anthropological experience in order to discern analogies, which may better clarify contemporary categories of thought.

We speak of the essence of the human person, thereby referring to the common manner through which every person shares in life, in being. We say, then, that the human person is a being that walks uprightly, laughs, thinks, creates, and possesses imagination, judgment, willpower, the ability to love, and so on. All these are characteristics of a common human essence or nature.

Naturally, human essence, this universal manner of human existence, cannot exist independent of individual and particular beings. Each individual being realizes the common essence in an actual expression; each person "hypostasizes"[4] the common essence, it constitutes the hypostasis of that essence. Human essence only exists within human hypostases.

Yet, the common features of essence, which are hypostasized by each individual person, are truly existent possibilities, characteristics of the manner in which every human existence is actualized. Here, we encounter the notion of the energies or actualization of essence or nature. Each human existence is potentially energized or actualized through its material and spiritual functions, in the expression of its reason, will, imagination, judgment, and so forth.

---

1. Gregory (335–394) was the younger brother of Basil the Great (330–379) and a mystical philosopher.

2. Maximus (580–662) was a monk, scholar, and theologian.

3. Gregory (1296–1359) was Archbishop of Thessalonika and defender of the hesychast monks of Mt. Athos.

4. "Hypostasis" is a philosophical term, meaning "substance" or "existence." It was the center of debate and controversy in the early councils of the church, which clarified and defined it as referring to the personal existence of the three persons of the Holy Trinity.

AN ORTHODOX ANTHROPOLOGY

Each human individual hypostasizes the common energies of human nature in a unique and unrepeatable manner. Each person has reason, will, imagination and judgment; yet each person reasons, wills, imagines, and judges in a unique, different, and unrepeatable manner. Consequently, the energies of human nature constitute an ontological reality, not simply because they characterize the common manner of existence among people but also because they express each person's hypostatic manner of being. The human energies, which are hypostasized within each person, also constitute and reveal the absolute existential otherness and uniqueness of each person.

We come to know a human person, the otherness of that person's existence, by means of the energies through which that person's being is realized and revealed. Therefore, we come to know the composer Johann Sebastian Bach by listening to his music; we come to know the artist Rembrandt by means of his paintings. The musical notes of Bach and the paint colors of Rembrandt differ in essence from the human essence of the two artists. Yet, the creative energy of one artist, which reveals the hypostatic otherness and uniqueness of that artist, is also actualized through different essences. The music, colors, writing, marble and clay actualize the *logos* (meaning or purpose) of the musician, artist, writer and sculptor. All of these reveal the personality of the artist, his or her existential identity and otherness.

The ontological content of this category of hypostasized energies permits us to attribute the ontological "principle" or beginning of the material world to a personal God. We do not attribute this to the divine essence (which would lead to pantheism), but to the hypostatic divine energies. The divine energies reveal the *logos* (meaning or purpose) of God's personal otherness in relation to the world, a meaning that is actualized in matter, yet that remains entirely different to God in essence.

Almost fifteen centuries before the quantum theory was conceived in contemporary physics, Greek ecclesiastical thought confirmed that matter is energy, "a syndrome of rational (or logical) attributes," the created result of the uncreated divine energies. The difference in essence between created and uncreated neither excludes nor hinders the created from being actualized as the *logos* (purpose or word) of the uncreated, from revealing the creative energy and hypostatic otherness of the personal divine *Logos* (or Word)—just as the notes and the colors, while being different in essence to their artists nevertheless actualize the *logos* (meaning or purpose) of the creative energy and hypostatic otherness of Bach and Rembrandt.

HUMAN BEINGS AND NATURE

It is only when people regard matter and nature in its entirety as a creation of a personal Creator that the use of matter and nature in its entirety constitutes a genuine relationship and not a uniform domination of humanity over the natural reality. It is only then that we are able to speak of an "ecological ethic," which derives its definitive character not from conventional rationalistic codes of conduct but from humanity's need to love and be loved in the context of a personal relationship. The *logos* (as "purpose" or "meaning") of the beauty of creation is an invitation addressed by God to humanity, an invitation to a personal relationship and communion with Him. It is a living and life-giving relationship. Contemporary ecology could then become the practical response of humanity to this divine invitation, a tangible participation in a relationship with God.

Is it, however, sufficient to speak of an ontological clarification of the meaning of matter and the world in order to provoke a different relationship between contemporary humanity and the material world? Surely not! In order for humanity to reach the point of responding to nature with the respect and awe that it responds to a personal artistic creation, this theoretical clarification must become an experiential knowledge and a social attitude. And, at this point, the role of social dynamics contained within the ecclesiastical tradition and community can be decisive. This will occur so long as the ecclesiastical conscience is purified from its estrangement into an ideological structure and from its inactive rest in the preservation of established forms.

The great challenge, which the Orthodox ecclesiastical conscience is called to appreciate today, is the surprising reality of the contemporary science of physics, the new fascinating cosmology that results from the study of quantum mechanics: namely, the potential of matter as energy, the relativity of space and time as the connection for the presence of matter, and the increasingly clearer anthropocentric purpose of the universe. The language of physics today reveals the universal reality as *logos* (purpose or meaning) that is actualized and only hypostasized in its encounter with the human personal *logos* (purpose or word).

If there exists a future for the demands of contemporary ecology, then this future is surely based, we would believe, in the free encounter of the historical experience in the Church of the living God with the experiential affirmation of God's *logos* (purpose or reason) actualized in nature.

Address at the Official Presentation of the Sophie Prize, Oslo, June 12, 2002

FREEDOM AND COST

All of our efforts to cultivate a sense of environmental responsibility and to promote genuine reconciliation among people demonstrate the immediate responsibility and initiative of the Ecumenical Patriarchate, which has served the truth of Christ for some seventeen centuries. Our Church regards the sensitization of its faithful in relation to the natural environment and in regard to the development of inter-religious dialogue as a central and essential part of its ministry of solidarity and co-existence.

We thank you wholeheartedly for the gracious invitation to address this auspicious occasion on one of the most critical global issues of our time: namely, the ecological crisis that we face. We still recall the recent news in regard to government representatives who did not come to an agreement about measures to be taken. This means that nations, which have the privilege of freely choosing their rulers, have not yet reached the point of environmental sensitivity demanded of their governments in regard to the cost involved. Therefore, having voluntarily assumed the effort of sensitizing people's conscience in the face of this crisis, we readily admit how much work there still remains.

THE BEAUTY OF THE WORLD

First of all, we must stress that it is not any fear of impending disasters that obliges us to assume such initiatives. Rather, it is the recognition of the harmony that should exist between our attitudes and actions on the one hand, and the laws of nature, which govern the universe, on the other hand. These laws have been established by the supreme personal Being, a Being that we call Trinitarian God that loves and is loved.

From the outset, we should state that we outrightly reject dualist opinions claiming that the world is the creation of evil, and is consequently evil. Furthermore, we also reject those opinions supporting the notion that material creation pre-existed and was simply fashioned by God, or the choice of others to believe that the body is the prison of the soul, which seeks to be liberated from the bonds of the former. Finally, we reject any opinions that demote humanity into a fragmented part of our earthly ecosystem, rendering it equivalent to every other part, and undeserving of any greater protection than that afforded to other species.

It is our conviction—and the truth of our conviction has been experientially confirmed—that both the material and spiritual worlds, visible and invisible things alike, are, according to Scripture and the Nicene-Constantinopolitan Symbol of our Faith, a "very good" creation by the good and loving God. On the basis, then, of such a conviction, we are able to articulate the fundamental principles of our worldview.

AN ENVIRONMENTAL CREED

We believe that the human person constitutes *the crown of creation*, endowed with the sacred features of self-consciousness, freedom, love, knowledge, and will. Such a teaching is part and parcel of our creation "according to the image and likeness of God."

We believe that the natural creation is *a gift from God to the world*, entrusted to humanity as its governor, provider, steward, and priest, in accordance with the commandments "to work and keep it," as well as to abstain from it partially. In this way, we admit the limitations as well as the responsibilities of humanity with regard to the natural environment.

We believe that the universe comprises *a single harmony or "cosmos,"* according to the classical Greek significance of this term, which implies a harmonious coordination of human will and human action on the basis of natural and spiritual laws established by the discerning, loving, and perfecting will of the divine Word.

We believe that humanity did not wish to coordinate *personal will and universal harmony*, in accordance with the divine plan. Instead, it preferred to pursue independence, resulting in the creation of a new order and different pattern within the natural environment—commonly referred to as anthropocentrism, but more properly identified as anthropomonism.

We believe that *a New Man, the God-Man, Jesus Christ,* appeared in the world, demonstrating perfect obedience to the original plan of the Father with regard to the relationship between humanity and the world. Jesus Christ reconciled the world to the Father. Henceforth, the world functions harmoniously through Him and in Him. He commanded us to use the world's resources in a spirit of ascetic restraint and eucharistic sacrifice, to transform our way of thinking from egocentrism to altruism in light of the ultimate end of the world. In the Greek language, again, the word for "end" (Greek: *telos*) implies both conclusion and purpose.

These brief principles describe our attitude and concern for the natural environment. We are endowed with freedom and responsibility; all of us, therefore, bear the consequences of our choices in our use or abuse of the natural environment. Yet, we also have the capacity to repent and the ability to reduce the damage of our actions in the world. We know, however, that the complete reconciliation and ultimate recapitulation of the world can only occur through Jesus Christ at the end of time.

Until then, God's unceasing love allows us only glimpses of that total reconciliation, to which we partially contribute when we abandon the abusive violation of nature and to accept it as a divine gift of love, treating it reasonably, gratefully, and fruitfully. Such is our dutiful response to the loving Creator, as well as to all those with whom we share this divine gift.

To imagine a world that functions in beauty and harmony, balance and purpose, in accordance with the overflowing love of God, is to cry out in wonder with the Psalmist, "How great are Your works, O Lord; You have fashioned all things in wisdom" (Ps. 86.10).

ORIGINAL SIN AND THE ENVIRONMENT

Our original privilege and calling as human beings lies precisely in our ability to appreciate the world as God's gift to us. And our original sin with regard to the natural environment lies—not in any legalistic transgression, but—precisely in our refusal to accept the world as a sacrament of communion with God and neighbor. We have been endowed with a passion for knowledge and wisdom, which open before us boundless worlds of the microcosm and the macrocosm, and present us with boundless challenges of creative action and intervention.

The arrogance that destroyed the Tower of Babel, through the misuse of power and knowledge, always lurks as a temptation. The natural energy wrought by the sun as a blessing on the earth can prove perilous when profaned by the hands of irresponsible scientists. The interventions of geneticists, which arouse enthusiasm in their potential, have not been exhaustively explored with a view to their side effects.

We are not opposed to knowledge but we underline the importance of proceeding with discernment. We also stress the possible dangers of premature intervention, which may lead, as Euripides emphasized, to "the desire to become greater than the gods,"[5] which the classical Greeks described as "*hubris*." Such discord destroys the inner harmony that characterizes the beauty and glory of the world, which St. Maximus the Confessor called "a cosmic liturgy."

ALL-EMBRACING LOVE

Our prayer and purpose join the priest in the Divine Liturgy, who chants the words during the *anaphora*: "In offering to Thee, Thine own [gifts] from Thine own, in all and through all—we praise Thee, we bless Thee, and we give thanks to Thee, O God." Then, we are able to embrace all—not with fear or necessity, but with love and joy. Then, we care for the plants and for the animals, for the trees and for the rivers, for the mountains and for the seas, for all human beings and for the whole natural environment. Then, we discover joy—rather than inflicting sorrow—in our life and in our world. Then, we are creating instruments of life and not tools of death. Then, creation on the one hand and humanity on the other hand, the one that encompasses and the one that is encompassed, cooperate and correspond. Then, they are no longer in contradiction or in conflict. Then, just as humanity offers creation in an act of priestly service and sacrifice to God, so also does creation offer itself in return as a gift to humanity. Then, everything becomes exchange, abundance, and a fulfillment of love.

Official Ceremony of the Binding Foundation, Vaduz, Liechtenstein, December 6, 2002

THE NEED FOR VIGILANCE

Your Excellency Prince Hans Adam II von und zu Liechtenstein, Beloved Guests: It is with great joy and emotion that we accept the honor of the

---

5. Euripedes (480–406 BC) was the last of the great classical Greek tragedians.

conferral on our Modesty of the award for nature and the protection of the environment.[6] Our joy is derived not so much from personal satisfaction for the recognition of our endeavors, as from the realization that these efforts have objectively contributed to alerting a good number of our fellow citizens throughout the world to the problem of the continuing underrating of the environment in our times.

The need to alert the whole of humanity to this matter is not only a personal choice of our Modesty, but also a conviction of the entire Orthodox Church. Already since 1989, the Holy and Sacred Synod of the Ecumenical Patriarchate, following a proposal of our predecessor Patriarch Dimitrios of blessed memory, designated the first of September of each year for Orthodox Christians, which coincides with the commencement of the Byzantine and ecclesiastical year of the Orthodox Church, as a day of prayer for the protection of the natural environment.

Our interest in the environment originates from our faith in its Creator and from our love toward Him and toward the fellow human beings that live within this environment. We believe in what is recorded in the first chapter of Genesis, the first book of the Old Testament, that God created all things "very good" (Gen. 1.31). Consequently, the natural environment and man, its ruler, are very good creations of God and their coexistence ought to be, according to God's plan, harmonic and mutually supportive, and not antagonistic and destructive. This is also the meaning of the commandment of God to the first created, Adam and Eve, to "work and to keep" the earthly garden. To work and to keep means nothing else than to retrieve from the natural environment reasonably and proportionally the necessary resources for human life and, simultaneously, to provide the necessary care for the environment so that it remains productive in the future in accordance with the plan of its Creator.

Typical of this point is what is written in the deutero-canonical book of the Old Testament, the Wisdom of Solomon, that, "God did not create death, nor does God delight in the loss of living beings. For he created all things in order to be, and the creation of the world is tied to salvation, and there is no poison of destruction in it, nor is there a kingdom of Hades on earth" (1.13–14). Then, the author of this book goes on to state

---

6. On the occasion of the presentation of the Binding Award "for nature and the protection of the environment."

that the impious people "invited death" (1.16) and to advise us all "not to be zealous for death by being deceived in your life" (1.12). The death, which the author advises us not to be zealous of, is not only that which occurs suddenly, but also the one that accompanies the corruption of the natural environment that sustains us.

THE BIBLE AND THE FATHERS

The reason behind the connection between corruption and death is that all things in the universe are interdependent and the destruction of one incurs a consequent damage for the other. This point is not a recent discovery of the environmentalists who like to speak about the balance of ecosystems. Athanasius the Great, Archbishop of Alexandria, had already written in the fourth century: "If one were to take the parts of the creation in themselves and consider each of them in his mind, he would certainly find that none of them is self-sufficient, but all of them are in need of each other and are constituted by the support of each other."[7] St. Gregory of Nyssa also writes in the same manner: "The whole universe has continuity in itself and there is no dissolution to the harmony of beings, but there is a concurrence of all with each other, and the entire universe is not torn apart from its entire co-inherence."[8]

Thus, working for the natural environment, we work for the harmony and balance of the universe and of our fellow human beings dwelling in it. Indeed, this concurs with the divine commandment, which we received in the person of our first created ancestors, that "we work and keep" the world, for we were placed in it by God as rulers to "have dominion over it," not of course as avaricious and greedy exploiters, but as rational and responsible partakers for the purpose of covering our real needs rather than amassing superfluous items.

It is certainly true that in each particle of nature there is a program, so to speak, that has been placed in it by God, a program that pertains to its development and role according to its participation in the universal harmony. Nevertheless, the supreme role of the coordinator and steward of earthly things (at least according to the data that are known thus far) was

---

7. PG 25, 53B.
8. PG 44, 724D.

reserved by God for the human being who had been endowed with peculiar intelligence according to His image. This is, in any case, the meaning of the allocation to man of working and keeping the earthly creation. Without the duty of preserving it, working it and keeping it are unthinkable.

THE DIVINE COMMAND TO CARE

Thus, being interested in the protection of the environment, we do not do anything more than observe the divine command that was given to us. And inviting our fellow human beings to be alert to their inherent duty, we fulfill our duty as religious shepherds and leaders, and as human beings that love all others and are concerned for the terms of our cohabitation on earth. This is because, according to our Orthodox faith, as it is specifically and characteristically presented by James, the brother of the Lord in his epistle: "Pure and undefiled religion before God the Father is this, to visit the orphans and the widows in their sorrow" (1.27).

In other words, this means to visit and to take care of everyone who is in sorrow, whether he is living in this planet now or whether he is going to live in it in the future as God ordained him to live. We certainly believe in the continuation of human life beyond the biological bodily death, but we know from the Gospel parable of the future judgment that the road leading to the life that lies beyond is the road of love for our fellow human beings and for God, which is expressed in practice through the observance of His commandments. One of these commandments, which is indeed among the first and was given before the disobedience of the created humans, is "to work and to keep" the creation of God for the sake of the whole of humanity, whether existing in the present or in the future.

We thank the Binding Foundation for awarding our endeavors in this area and all who honored our Modesty by their presence at this ceremony. We also wish everyone all the best, and for humanity, that it may come to understand that the road of irrational and avaricious use of earthly resources and their destruction for a variety of reasons is detrimental to all of us and thus we all must undertake our responsibilities.

RELIGION, SCIENCE, AND THE ENVIRONMENT SYMPOSIA

Address at the Opening Ceremony of Symposium II, Trabzon,
Turkey, September 20, 1997

CREATOR AND CREATION

Among many prevailing misconceptions is the idea that religion in general, and Christianity in particular, are interested only in life after death, as if by struggling to divest one's self of a material body, one might be changed into a liberated spirit, as Plato taught. However, leaving the arguments of other religious teachings and philosophical schools to their own adherents, we will focus here on the Eastern Orthodox Church, in recognition of our responsibility.

It is a qualitative element in our faith that we believe and accept that the Creator, who loves us, fashioned humankind in its psychosomatic reality and also fashioned the material universe according to His image and likeness (Gen. 1.26), making it and us "very good" (Gen. 1.31). Consequently, we stand far from any dialectical opposition of matter and spirit, body and soul, or the present world and metaphysical reality. We reject this Manichean perception and live in harmony with that which we are, in harmony with the material and spiritual world within which we live and move and have our being. We live in harmony with God, the Creator of all things and with all things, which He has created.

Obviously, we recognize a certain hierarchy in the created order and we accept a gradation of proportionate value in the harmony of creation, inasmuch as the omniscient Creator has arranged all of creation in the fullness of His love. Even when we cannot know the specific reason of creation and the existence of an objective reality, or even the property of a physical law, we can still be absolutely certain of one supreme *Logos*, who has established and apportioned all things with immanent rationality. It is for this reason that we hesitate to opine that human remedies, which interfere with the natural order, are better than those found in the natural conditions provided by God. However, it does not follow from this that we consider the natural order to be an intangible "taboo," for we possess the commandment of God that we should have dominion over the earth.

ENVIRONMENT AS HOME

We should focus our attention from the outset on the proper limits of our interventions, such as they are, and around a consideration of what might be considered appropriate interventions. We can speak in a positive manner and without hesitation about these things. In describing the appropriate limits of our interventions in the natural condition of the material world, in each particular instance we stand on a razor's edge, which itself is a measure of our own spiritual condition. Our answer presupposes a careful weighing of innumerable factors, which must take place with prudence and dispassionate discernment. Unfortunately, this is rare, much like our own virtue is rare.

The first part of the word "ecology" is derived from the Greek word "*oikos*," meaning our "home." There is more to our "home" than people, land, water, plants and animals. That is the spirit of the term. Yet, people rarely take time to look at the beauty of this world or to feel the spiritual connection between our surroundings and us. In order to change our behavior toward the "home" we all share, we must rediscover spiritual links that may have been lost, and reassert human values in our behavior.

HUMANITY AND NATURE

And so we stand before nature, which has in recent times come to be called the "environment," with the same affirmation of its own worth as that with which the Creator of all things characterized it as "very good." However, at the same time, we face the reality of the hierarchy inherent in the design of the cosmic plenitude, both the material and the spiritual, and we assign it to its proper place. We do not make of nature an absolute value, nor do we deny it to the point of absolute negation. Nature exists for the service of humanity according to the material portion of our nature; it is not humanity that exists for the sake of nature.

With this understanding, the core of our concerns for the Black Sea and of the environmental dangers threatening it is focused on those people who live in the surrounding regions of the Black Sea, and not on the Black Sea in itself. We are interested in preventing a catastrophe of the Black Sea as a biotope. Therefore, we declare that our most vivid concern is for our fellow human beings who dwell either near or far from it and who may not be aware of the danger that threatens their life, health, and livelihood through the disruption of the natural ecological balance of the Black Sea.

The Black Sea is a remarkable place with a long history, both natural and human. Unlike other seas on our planet, it was formed comparatively recently, when the waters of the Mediterranean broke through the Bosphorus[9] valley, flooding a vast low-lying freshwater lake, some seven or eight thousand years ago. People living in primitive settlements will have watched this great event as a salty cascade that surged down the valley to give new life to the new sea. Subsequent generations have admired the beauty of this sea, enjoyed the fish it provided, and sailed its waters. Different civilizations have come and gone, but always the sea seemed an immutable part of eternity, governed by the "everlasting covenant between God and every living creature upon earth" that was proclaimed in the book of Genesis after Noah and his Ark survived the mighty floodwaters. Now, almost as dramatically as it was formed, the people are beginning to see change: fish are disappearing, crystalline waters are becoming choked with green algae, beaches and harbors are badly polluted, and the Black Sea's beauty is soiled by the hand of humankind.

We are deeply concerned for their daily life, for it, too, is a gift of God, "very good" indeed, and destined to be used as a starting-point for an eternal life, continuing beyond the grave into the infinite, when the bodily substance of our nature is transmuted into a spiritual reality.

ASCESIS AND CONSUMPTION

To ensure the furtherance of natural life, we consider ascetic self-denial to be necessary and the reduction of many material pleasures to be beneficial. However, such deprivation should certainly be a voluntary and conscious decision, and not something imposed by external constraints and consequently without ethical merit. Therefore, from this perspective, we join forces with all those who are concerned for the self-sufficiency of those material goods necessary for survival, but we do not approve of overproduction, irrational exploitation, and greed. We seek and urge the patristic and Aristotelian sense of moderation and balance and would dissuade you as

---

9. The Istanbul strait that marks the boundary between Europe and Asia.

much from forming an idol of wealth as from reducing wealth to an evil in itself. For, wealth is the accumulation of the abundance of human labor and natural resources. Wealth is both valued and reviled in an ethical sense, according to the manner of its concentration and the manner of its use.

It is obvious that the rapacious exploitation of natural resources, which derives from greed and not from an extraordinary circumstance of need, entails a debilitation of nature to the extent that it cannot renew its own productive powers. Such exploitation is as unethical as it is irrational, given that, as is observed sociologically, a lack of ethics coincides with a lack of intelligence to an astounding degree because someone who is truly intelligent understands the injurious nature of bad character. Again, it is equally evident that the utilization of concentrated wealth for selfish purposes, for one end or another, for good or bad, is ethically disdainful. Rather, the utilization of the concentrated surplus of human effort and natural resources forms a basis of the furtherance of the means of production, aimed at the improvement of the quality of life in human society and for the allocation of various aspects of education, public health, and other possibilities for the development and enrichment of the spiritual life. This justifies wealth as a shared property, expended for the common good, even if it is managed by only one part of human society.

The Orthodox point of view and teaching is that the environment was created so that we might evolve and develop within it in body and soul. In addition, we ought to preserve the divine law and do those things that are useful and reasonable for our self-sufficiency and not for any kind of hoarding. If there should happen to be any excess, without draining the vigor of nature dry, this ought to be employed again for the common good. Thus, religion propounds eternal truths and maintains sensitivity to all members of society because of the danger of falling away from those same truths.

SCIENCE, POLITICS, AND PEOPLE

It is at this point that the work of science begins, for science certifies the situation, proposes measures that would delay the worsening of this situation, and organizes the effective application of these measures. Not infrequently, whenever science considers it possible through the measures at hand, it also proposes the transformation of the destroyed elements in the environment.

This symposium has the blessing of gathering, as its title declares, representatives of religion and science. All of you enjoy the blessings of the Holy and Great Church of Christ, and personally speaking, most certainly the love and the prayers of our Modesty; you enjoy the patronage of the President of the European Commission, His Excellency Jacques Santer.[10] The subject of this symposium is of interest to many peoples and their governments. The circumnavigation of all those interested countries as the venue for this symposium constitutes a conscious decision, aimed at sensitizing individuals and whole peoples to the significance of the biotope of the marine region of the Black Sea for their life. Environmental change and destruction on such a vast sea-wide scale, through the thoughtless contribution of many, wreaks vengeance over a long period of time and in all directions. The redress of such circumstances is a most onerous task, requiring the cooperation of all concerned parties. That is why it constitutes a challenge to the ethical and social conscience of us all and substantiates a criterion of the degree of our spiritual condition.

The Black Sea is spoiled and its biotope destroyed because of the indifference of countless people, each of whom contributes to a greater or lesser extent to the unfortunate result. A ship's captain, for example, protected by the fact that it is not easy to prove that he discards his ship's polluted waste into the sea, is as responsible as the person who lives far from the coastline, but who discards personal destructive waste into the river, which empties into the Black Sea. Unless everyone is made sensitive to the harmful nature of such actions, it is almost impossible for any endeavor to improve the situation.

The Role of Religion

It is precisely here that the fundamental role of religion is revealed. Religion can inspire the behavior of every individual, in turn motivating more individuals and creating mass movements, which can spread the necessity and benefit of these behaviors.

The fact is that the negative results spring from the contribution of many people, who may hardly be aware of their own responsibility, because their contribution to the overall result is relatively minor. If there is

10. The ninth president of the European Commission (1995–1999), Santer was previously prime minister of Luxembourg (1984–1995).

going to be an overturning of this problem, it will require the cooperation of everybody; in other words, the solution will reveal the necessity of achieving a social ethic and a common effort. There is no single deed that benefits others, without it first benefiting us.

This sense of a common fate, the polar opposite of the widespread individualistic and self-interested perception, which is shortsighted in its appreciation of the world, is a basic teaching of the altruistic Christian faith, and especially of Orthodoxy. Without it, facing problems such as those of the Black Sea would be impossible.

Surely there are not many people today who are ignorant of the influence of prevailing ideas and beliefs on our lives. To bring new ideas to predominance, we must focus on the acquisition of adherents and on the spreading of ideas. With this understanding, we seek also through this symposium to activate for the common good the dormant sense of responsibility, which exists among every people and in each individual. In this way, the aim of a healthier Black Sea may be more easily achieved for the benefit of all humankind.

We are deeply pained by the suffering, which, because of human indifference, keeps piling on our fellow human beings. We call on every conscience to awaken! We invite those of our beloved fellow participants whose consciences are already sensitized to a virtual apostolic commission, to disseminate ideas about the necessity for a common confrontation of these problems.

Address at the Opening Ceremony of Symposium III, Passau, Germany, October 17, 1999

DANUBE—RIVER OF LIFE

The international symposium on the subject, "Danube—A River of Life," forms part of the broader program Religion, Science, and Environment, and also the continuation of our symposium held two years ago on the subject: "The Black Sea in Crisis." We should, therefore, like to address first of all my heartfelt greetings and warmest welcome to our dear participants and all others attending this inaugural session. We also feel a genuine obligation to address wholehearted thanks to the ecclesiastical, municipal, and political authorities of this historical and most picturesque city of Passau, which is offering us its hospitality. Our special thanks go to the Most Reverend Dr. Eder, Bishop of Passau, the Most Honorable Lord Mayor of the City, and all those who have contributed to our hospitality and to the flawlessly organized inauguration of the symposium here.

Passau is built at the confluence of the Inn, Ilz, and Danube rivers, and is situated on the frontier between Austria and Bavaria, now one of the federated states of Germany. Thus, it constitutes an ethnological and physical frontier point, as from this point on, traveling on the Danube becomes easier. However, although frontiers serve in general as points of separation, it seems on the contrary that this city has had the historical mission of unification. As we all know, it was here that, in the year 1552, peace was reached between Roman Catholic and Evangelical Protestant Germans, while its significance as a center of trade underlines the peaceful cooperation between peoples, a cooperation that benefits all of them, as is the case with every healthy commercial activity.

The Danube itself is, as we know, a natural frontier that divides peoples and states into many sections. It is also a water route that unites and brings into contact all those whom it separates and, through them, even those who live beyond its banks. However, in addition to its attribute as a communications artery, which, through international treaties, has been declared free for shipping, it is also a bearer of enormous quantities of water from continental Europe toward the Black Sea, as well as of mud and toxic wastes. This is why the Danube is in danger of becoming, in the future, an easy dumping ground for pollutants and toxic substances from areas along its banks and carrying them to the Black Sea.

It is well known that the Danube empties 6,000 to 7,000 cubic meters of water per second into the Black Sea. From this standpoint it is a very important supplier of water to this Sea, helping to revitalize it. However, the growth of industry in all the regions along the banks of the river, as well as in regions of the interior that communicate with the Danube by means of tributaries, has transformed the Black Sea into a second Dead Sea, as was pointed out during the symposium of 1997.

HUMAN IMPACT AND INFLUENCE

Hence, it is the duty of all persons of good faith to contribute, each according to his or her position and ability, not only to keeping the

Danube a river with free shipping and transportation of goods from one region to another, but also to preserving it free from toxic substances, in order to maintain the ecological equilibrium of the physical areas surrounding it and the Black Sea.

At this point, I leave: we cannot fail to express our deepest distress at the recent bombing of Yugoslavia, which also affected the waters of the Danube and caused ecological damage that will be difficult to correct. The expression of my distress is not, of course, confined exclusively to the ecological effects of the bombing. For, I leave: we have repeatedly expressed our deepest sorrow over the fact that this bombing was responsible for killing people; destroying cultural monuments, monasteries, and churches; incapacitating factories; contaminating human communities, animals, and plants and nature in general with toxic substances; and generally causing material and moral damage on a massive scale. However, the object of the present international scientific symposium obliges us to focus our interest on the ecological consequences of this bombing and to point out that its adverse effects are manifested not only in regions neighboring the areas directly affected, but also in regions throughout the length of the Danube as well as those lying on the Black Sea coast.

Of course, we could say, without going beyond the bounds of realism that human acts have repercussions far beyond the limits of these acts, which we usually regard as extreme in each case. There is a known scientific view according to which the world's physical equilibrium may be disturbed through the flight of a butterfly in Japan, which through some chain of events causes rain in America. This means that we never know exactly when or how, but we do know with certainty that the pollution of our environment will also have consequences for the polluters themselves, regardless of how far away they are from the point at which the pollutants have been dumped. The fact that the effects of pollution also travel upstream from the direction in which the river waters flow must make us all think.

SENSITIVITY AND FORESIGHT

At this point, allow us to remind you of certain details that we have received from Herodotus, according to whom there existed in his time a people that considered rivers sacred and polluting them to be sacrilege. Perhaps those who demythologize ancient beliefs may regard such faith as superstitious. However, even this superstition is socially preferable to the unscrupulous and irresponsible dumping of harmful substances into the rivers, temporarily relieving those who selfishly pollute the river, but substantially harming the next generation of their fellow human beings who are going to use it.

Therefore, we must acquire a moral code higher than the one used by such crude people and learn to respect humanity, accepting as a basic principle of our behavior that it is morally unacceptable to burden others with our wastes. This is the only way to help ensure that the Danube, the longest river of this region—as well as the other nearby rivers—becomes a road of life for all. Otherwise, all of these rivers will end up being bearers of death, a death that is sown by many selfish people to the detriment of their fellow human beings and nature as a whole.

This is the deeper reason why our Modesty, whose primary mission is the Christian education and sanctification of the Orthodox faithful, has wholeheartedly adopted the present series of international ecological symposia. The reason is that, as the Church Fathers also teach, the root of all evils that plague humankind is selfishness, and the highest expression of virtue is selfless love. It is, therefore, not permissible for a faithful Christian who is seeking sanctification to remain indifferent to the effects of his or her acts on their fellow human beings. The sensitivity of their conscience must be increased so that they are not indifferent even to the indirect consequences of his acts. As Abba Isaac the Syrian observes, the sensitive and charitable heart "cannot stand even to hear of sorrow, even the slightest such sorrow, in creation. That is why the heart of such a person grieves for the creatures not endowed with reason, as well as for enemies of the truth and even for those who harm one. Such a person addresses to God a prayer for them in tears, that God may spare them and have mercy on them; and, similarly, one prays for the reptiles, because one's heart is full of mercy, similar to the mercy that fills the heart of God."[11] This saintly sensitivity is, of course, possessed by very few. But this does not mean that we should go to the other extreme, to the complete lack of sensitivity. For, as St. John Climacus[12] says: "The hardened person is a foolish philosopher."[13]

---

11. *Ascetic Treatises*, Homily 48.

12. John Climacus (579–659), author of *The Ladder of Divine Ascent*, was abbot of the monastery on Mt. Sinai.

13. *Ladder of Divine Ascent*, Step 17.

In the context of our pastoral concern to raise everybody's awareness to the effects of ecological catastrophe on humankind, we have been active in a variety of directions, among which is participation in the present symposium and the organization of the annual seminars and an educational Ecological Institute at Halki. In all these, the altruistic importance of protecting the environment is emphasized, together with the pleasure God takes in it. In other words, it is stressed that our motives are not simply inspired by our love for nature—although those who have such inspiration have nobler motives than those who are indifferent to nature and to its destruction—but are strictly human-centered, just as in fact all of creation is anthropocentric. Our Christian faith teaches us—and the modem scientists also accept—that the world was created for the sake of humankind and that everything is regulated so as to contribute to our survival. Of course, we live in a world that is partly deregulated, but this is the consequence of our revolt against the harmony of God, which brought with it a partial revolt of nature against our rule over it.

ECOLOGICAL CURSE AND DIVINE PROVIDENCE

It is known that the Old Testament words "Cursed is the ground because of you thorns and thistles it shall bring forth to you" (Gen. 3.17–18) were spoken by God to Adam after his disobedience; they express, according to the Law of God, the relationship between an act and the consequences of the act. Nonetheless, these words have given rise to much misunderstanding concerning the goodness of God and the condition of the natural world. At first sight, God gives the impression that He is cursing creation. Yet, if we look deeper into the spirit of the Holy Scripture, especially as our Lord Jesus Christ in the New Testament developed it, we realize that this was not a punishment imposed by God on the transgressor. Rather, it is God's communication to humanity about the new ontological interaction, which was created because of our disobedience. It was not, in other words, a legal connection between transgression and punishment. Rather, it was an ontological relationship between a situation where man was in communication with God and accepted the favorable influence of divine grace and love on him and the world, and a new situation where man tried to become independent of God, rejecting communion with Him and the influence of divine grace, which this love created on him and on creation. From that point onward, humanity was therefore evolving independently, refusing the divine grace in the midst of a world whose initial harmony was ontologically altered.

"The love of God never fails and never will fail" (1 Cor. 13.8). Yet, humanity rejected God's love and its gifts and tried to create an independent path. As Abba Isaac says: "Those who feel that they have failed in love" feel such sorrow in their hearts as though they were in the midst of the worst hell. "For the sorrow that hits the heart due to failing in love is sharper than the one that results from any hell."[14]

The love of God never fails. And the new situation, which resulted in the world because of Adam's refusal to comply with God's loving plan, namely the plan of developing a perfect and constantly improving loving relation between God and humanity, was disagreeable only because it fell short of love, because love is that personal relationship that offers true beatitude and happiness. Because of the love of God, Adam was created with a dual nature, composed of matter and spirit; and because of this love of God, Adam was given an opportunity to enjoy the infinite treasure of God's material and spiritual gifts.

God's love never abandoned humanity, even when humanity followed its own independent path. God's love has given to us nature, good in itself—namely, this earthly and broader cosmic space in which we live—so that from this space we may draw everything we need to live, and, at the same time, we can practice virtue, a part of virtue being the use of nature with reason and gratitude. Unfortunately, some people do not use creation in a moderate way, but abuse it and create problems for others and for the harmony of the natural functions in the physical world that remained in place after the Fall. Although nature changed ontologically after the Fall, nevertheless it has not ceased to function according to the divine plan and the divine laws embodied within it.

HUMAN RESPONSE AND ASCETIC STRUGGLE

Within the framework of this rebellious and partly dysfunctional nature, humanity must fight both for its physical survival and for its spiritual improvement. "All our life must be a struggle full of pain," St. John

---

14. *Ascetic Treatises*, Homily 84.

Chrysostom warns us,[15] as do all the Church Fathers. Yet, the purpose of this struggle is not to strip us of our physical body, which some philosophers have wrongly considered the prison of our soul. Instead, it is to lead to the Aristotelian middle road of developing the spirit and preserving the body, perceived as a useful tool for humanity as a whole, under the guidance of the mind governing it. Christians living on earth and having their city in heaven[16] reject neither earth nor heaven. At the same time, however, they do not feel inseparable from earthly goods and they do not completely reject the earth, having espoused, as they say, the heavenly life. Considering this life as a place of exercise and not as one of one-sided resignation, with the purpose of being killers of selfish passion rather than killers of the body—as one early ascetic once told someone who tried to become perfect through excessively ascetic discipline. They have always before their eyes what the Apostle Paul wrote to Timothy: "Every creature of God is good, and nothing is to be refused if it is received with thanksgiving. For, it is sanctified by the word of God and prayer" (1 Tim. 4.4).

Thus, according to St. John Chrysostom, our struggle in the present life consists of becoming virtuous in the right way, so that God, in His great glory, will raise our bodies.[17] Life thus becomes a joyous and creative struggle, full of good works and acceptance of the whole of the creation of God as very good. That is why the Church prays continuously for the success of all good human works, for the blowing of favorable winds and for the rich harvest of the fruits of earth. And it expresses continually its admiration for the beauty of God's creation, in a common hymn with the Psalmist: "Oh Lord, how manifold are your works; in wisdom you have made them all" (Ps. 104.24).

It is, therefore, in this atmosphere of exuberance, joy, thanksgiving, and creativity that the Christian practices a rational use of material goods and natural resources, always retaining a sentiment of awe and sensitivity toward the whole of creation. The Christian resembles a wise administrator of the commandment to work and keep the earth capable of providing humanity with food and pleasure for the glory of the all-wise and benevolent Creator.

---

15. PG 47.453. Born in Antioch, St. John Chrysostom (347–407) served as Archbishop of Constantinople and was one of the most popular preachers in Christendom.

16. See the *Letter to Diognetus*, PG 2.1173.

17. PG 54.636.

In spite of the disobedience, in spite of the ontological alteration of the world as a result of the interruption on our part of God's loving relationship, the Creator has never ceased to do everything for our sake and for the sake of readmitting us into the beatitude of love, the fullness of which constitutes the infinite perfection of happiness.

Therefore, the Church says "yes" to God's creation, while at the same time inviting everybody through this affirmation to reach out to the Creator and accept His invitation into a relationship of love. However, we do so in such a way that we do not enjoy the gifts of God as ungrateful recipients, but as grateful and noble receivers, expressing our thanks and our love to Him and thus helping to bring about the eternal and indestructible relation of love, which includes eternal life.

Having, then, before us the fact that our final goal is to serve the human race by maintaining our natural surroundings in a healthy state, we salute with satisfaction the symposium that has just begun. we wish with all our heart for every success in its proceedings and welcome our dear participants whom we know well and whose wise contributions we appreciate, thanking our hosts once more and invoking upon all the grace and eternal mercy of God.

*Address at the Opening Ceremony of Symposium IV, Durres,
Albania, June 6, 2002*

A SEA AT RISK, A UNITY OF PURPOSE

First, we thank God the Creator, Provider and Governor of the Universe, for the joy of this auspicious gathering, motivated by our common and vivid interest in the environment granted to us by the Creator. We address a heartfelt, friendly greeting and wish you all a pleasant stay, inspiring thoughts, creative deliberations, and a positive outcome in this sea-borne symposium, as well as a good harvest of profitable conclusions.

The present conference constitutes part of a series of similar symposia, during which the environment, and especially that which is connected with rivers and seas, is examined from the points of view of its present as well as its ideal conditions, based on both religious principles and scientific precepts. This year, our exploration probes more deeply into the environmental ethos that determines our attitude toward the environment.

Our capacity as the first among equal bishops of the Orthodox Christian Church obliges us to offer a brief exposition of our faith concerning the environmental ethos in order to contribute to the formulation of a commonly acceptable environmental ethos, which may serve as a guide to further action.

THE TERM "ENVIRONMENT"

It is a fact that the term "environment" presupposes someone encompassed by it. The two realities involved include, on the one hand, human beings as the ones encompassed, and, on the other hand, the natural creation as the one that encompasses. In our discussion, then, of our environmental ethos, we must clearly retain this distinction between nature as constituting the environment and humanity as encompassed by it.

We underline this point because it is widely held among certain ecologists that humanity is classified within the natural ecosystem, inasmuch as it is considered of equal significance with every other living being. This demotion of humanity, sometimes characterized as humility, constitutes a reaction against anthropocentric arrogance. This is due either to a complete rejection of God by human consciousness and the corollary divinization of humankind, or else to a misunderstanding of the divinely ordained relation of humanity to the rest of creation, whereby humanity lords unrestrainedly and abusively over it. In both cases, the attitude toward nature is criminal and destructive because humanity regards nature myopically and selfishly.

It is, therefore, appropriate and imperative to respond to and react against this attitude and ethos, especially when the wide-ranging and universally destructive consequences resulting from them have become the daily talk of the experts. Yet, it is not proper for the distinction between human beings and created nature (between the ones encompassed and the environment that encompasses) to be abolished in a way that equates humanity with the rest of the created beings.

The various proposals of this "deep ecology" have—whether it is admitted or not—no religious basis and lead to a passive attitude with regard to human disasters, which are explained as natural consequences of the ecological equilibrium. In this case, creation is rendered divine as a whole, and humanity's unique position therein is not recognized, except with regard to the Creator's action upon the environment.

BIBLICAL PRINCIPLES

The Orthodox Church assumes as its starting point the teaching of the Bible, accepted by the three great monotheistic religions, introducing a third factor in the relationship between humanity and the environment. This factor is the Creator of both humanity and the environment, who provides for all and has laid down the laws of harmonious coexistence of all elements in the universe, both animate and inanimate, endowing humankind with the mandate to serve as king of creation and the command to cultivate and preserve it. Cultivating the environment implies collecting from nature all that is required for our material survival and spiritual growth. Preserving the environment involves the obligation to respect this divine gift and not to destroy it in order to fulfill its initial purpose responsibly and reasonably.

Let us concentrate on certain elements of this biblical account of the divinely ordained relationship between humanity and the rest of creation. The first noteworthy point is the restriction placed on the first-created not to consume a certain fruit. Beyond serving as a basis for Christian asceticism, this commandment is a clear indication of environmental significance that the authority granted to humanity over nature is not absolute. While humanity was created to rule over the earth and all therein, it rules subject to restrictions and rules ordained by the Creator. Trespassing against these rules results in fatal consequences. The profound symbolism and extensive implications of this fact for an environmental ethos are apparent. Today we witness death approaching because of trespassing against limits that God placed in our proper use of creation.

A second noteworthy point is that the gift of the paradise of delight to the first-created was accompanied by the commandment and responsibility of humanity "to work in it and to preserve it" (Gen. 2.15). Working and preserving constitute a duty of active responsibility. Therefore, any principle of passivity or indifference toward environmental concerns cannot be regarded as sufficient or proper.

A third point, equally worthy of our attention, is that the consequences of the transgression of the first-created also had cosmic implications, rendering the earth cursed on account of their actions (see Ex. 3.17) and producing thorns and thistles in the environment. This rebellion incurred the gradual corruption and ultimate destruction of the ecological balance, which continues to this day, whenever we violate the commandment of preservation and abstinence, proceeding instead to misuse and abuse the earth.

Finally, we should observe that the Creator also took special care during the great flood, so that through Noah, the plants, the clean animals that were directly useful to humanity as well as the unclean ones that appeared to be of no consequence, should be preserved from extinction. This divine concern constitutes a clear vindication of our interest in the survival of those living species that tend toward extinction.

Contemporary scientific research underlines the wide-ranging environmental consequences of human behavior in a particular time and specific place. This constitutes an experiential recognition of the religious truth that Adam and Eve's act of transgression wrought an important change and a fatal corruption for the entire world. This is not the result of a legal-ethical responsibility bequeathed to future generations, but of an irreparable disruption in the natural harmony. In a musical symphony, a single note of dissonance can destroy the entire performance; the only possible remedy is the repetition of the concert. In ecclesiastical terminology, this is called "regeneration."

Roman Law recognized the absolute authority of a landlord over property. Yet, it stipulated certain restrictions concerning the disposal of unwanted materials, determined according to their usefulness and effectiveness. These principles should also govern the disposal of modern industrial wastes in relation to their detrimental impact on others.

Therefore, the entire universe constitutes one community and the actions of any single member affect every member of the community. The traditional Christian doctrine, both concerning the destructive evil committed by Adam, whereby corruption was introduced into the world, and the restorative good enacted by the new Adam, whereby new life was introduced into the world, provides a critical basis for the formulation of a new environmental ethos. Such an ethos is clearly warranted by the global impact of every behavior, both proper and improper.

THE WAY OF THE SAINTS

The saints have always taught that no one is saved alone and, therefore, that no one should strive for individual salvation but for the salvation of the whole world. Such a teaching is affirmed in the environmental field and confirmed by science. This conviction constitutes an essential aspect of the environmental ethos, required both of believers who rely on the precepts of faith and of those who wish to establish an ethos based on reason.

This concern for the salvation of all humanity and the preservation of all creation is translated into a merciful heart and sensitive attitude, so characteristically described by the seventh-century ascetic, Abba Isaac the Syrian. We are responsible not only for our actions, but also for the consequences of our interventions. After all, no responsible ruler leaves the growth of one's people unplanned and to the mercy of fate. Rather, a wise ruler assumes appropriate measures for the people's growth in accordance with specific goals. As ruler of creation, then, humanity is obliged to plan for its preservation and development. This requires the recruitment of scientific knowledge and involves the respect of all life, especially of the primacy of human life. It is precisely such a vision that also constitutes the fundamental criterion for any environmental ethos.

CONFERENCES AND EVENTS

Keynote Address at the Santa Barbara Symposium, California, November 8, 1997

A RICH HERITAGE

The Ecumenical Throne of Orthodoxy—in preserving and proclaiming the ancient patristic tradition and of the rich liturgical experience of our Orthodox Church—today renews its long-standing commitment to healing the environment. We have followed with great interest and sincere concern, the efforts to curb the destructive effects that human beings have wrought upon the natural world. We view with alarm the dangerous consequences of humanity's disregard for the survival of God's creation.

It is for this reason that our predecessor, the late Patriarch Dimitrios, of blessed memory, invited the whole world to offer, together with the Great Church of Christ, prayers of thanksgiving and supplications for the protection of the natural environment. Since 1989, every September 1st, the beginning of the ecclesiastical calendar, has been designated as a day of prayer for the protection of the environment throughout the Orthodox world.

Since that time, the Ecumenical Throne has organized an Inter-Orthodox Conference in Crete in 1991, and convened annual Ecological Seminars at the historic Monastery of the Holy Trinity on the island of Halki, as a way of discerning the spiritual roots and principles of the ecological crisis. In 1995, we sponsored a symposium, sailing the Aegean to the island of Patmos. The symposium on Revelation and the Environment, AD 95 to 1995, commemorated the 1900th anniversary of the recording of the Apocalypse. We have recently convened a transnational conference on the Black Sea ecological crisis, which included the participation of all nations that border the sea.

In these and other programs, we have sought to discover the measures that may be implemented by Orthodox Christians worldwide, as leaders desiring to contribute to the solution of this global problem. We believe that through our particular and unique liturgical and ascetic ethos, Orthodox spirituality may provide significant moral and ethical direction toward a new generation of awareness about the planet.

LITURGY AND LIFE

We believe that Orthodox liturgy and life hold tangible answers to the ultimate questions concerning salvation from corruptibility and death. The Eucharist is at the very center of our worship. And our sin toward the world, or the spiritual root of all our pollution, lies in our refusal to view life and the world as a sacrament of thanksgiving, and as a gift of constant communion with God on a global scale.

We envision a new awareness that is not mere philosophical posturing, but a tangible experience of a mystical nature. We believe that our first task is to raise the consciousness of adults who most use the resources and gifts of the planet. Ultimately, it is for our children that we must perceive our every action in the world as having a direct effect upon the future of the environment. At the heart of the relationship between man and environment is the relationship between human beings. As individuals, we live not only in vertical relationships to God, and horizontal relationships to one another, but also in a complex web of relationships that extend throughout our lives, our cultures, and the material world. Human beings and the environment form a seamless garment of existence, a complex fabric that we believe is fashioned by God.

People of all faith traditions praise the Divine, for they seek to understand their relationship to the cosmos. The entire universe participates in a celebration of life, which St. Maximus the Confessor described as a "cosmic liturgy." We see this cosmic liturgy in the symbiosis of life's rich biological complexities. These complex relationships draw attention to themselves in humanity's self-conscious awareness of the cosmos. As human beings, created "in the image and likeness of God" (Gen. 1.26), we are called to recognize this interdependence between our environment and ourselves.

In the bread and the wine of the Eucharist, as priests standing before the altar of the world, we offer the creation back to the Creator, in the context of a mutual relationship to Him and to each other. Indeed, in our liturgical life, we realize by anticipation, the final state of the cosmos in the kingdom of heaven. We celebrate the beauty of creation and consecrate the life of the world, returning it to God with thanks. We share the world in joy as a living mystical communion with the Divine. Thus it is that we offer the fullness of creation at the Eucharist, and receive it back as a blessing, as the living presence of God.

THE ASCETIC WAY

Moreover, there is also an ascetic element in our responsibility toward God's creation. This asceticism requires from us a voluntary restraint in order for us to live in harmony with our environment. *Asceticism offers practical examples of conservation.* By reducing our consumption—what in Orthodox theology we call *enkrateia* or self-control—we come to ensure that resources are also left for others in the world. As we shift our will, we can demonstrate a concern for the third world and developing nations. Our abundance of resources will be extended to include an abundance of equitable concern for others.

We must challenge ourselves to see our personal, spiritual attitudes in continuity with public policy. *Enkrateia* frees us of our self-centered neediness, that we may do good works for others. We do this out of a personal love for the natural world around us. We are called to work in humble harmony with creation and not in arrogant supremacy against it. *Asceticism provides an example whereby we may live simply.*

Asceticism is not a flight from society and the world, but a communal attitude of mind and way of life that leads to the respectful use, and not the abuse, of material goods. Excessive consumption may be understood to issue from a worldview of estrangement from self, from land, from life, and from God. Consuming the fruits of the earth unrestrained, we become consumed ourselves by avarice and greed. Excessive consumption leaves us emptied, out-of-touch with our deepest self. *Asceticism is a corrective practice, a way of* metanoia, *a vision of repentance.* Such a vision will lead us from repentance to return, the return to a world in which we give as well as take from creation.

REPENTANCE FOR THE PAST

We invite Orthodox Christians to engage in genuine repentance for the way in which we have behaved toward God, toward each other, and toward the world. We gently remind Orthodox Christians that the judgment of the world is in the hands of God. We are called to be stewards and reflections of God's love by example. Therefore, we proclaim the sanctity of all life, the entire creation being God's and reflecting His continuing will that life abound. We must love life so that others may see and know that it belongs to God. We must leave the judgment of our success to our Creator.

We lovingly suggest to all the people of the earth, that they seek to help one another to understand the myriad ways in which we are related to the earth and to one another. In this way, we may begin to repair the dislocation that many people experience in relation to creation.

We are of the deeply held belief that many human beings have come to behave as materialistic tyrants. Those that tyrannize the earth are themselves, sadly, tyrannized. Rather, we have been called by God to "be fruitful, increase and have dominion in the earth" (Gen 1.28). Dominion here is a type of the kingdom of heaven. Thus, St. Basil of Caesarea[18] describes the creation of humanity in paradise on the sixth day as resembling the arrival of a king in the palace. Dominion is not domination; it is an

---

18. Basil the Great (330–379) was a prominent bishop and influential theologian.

eschatological sign of the perfect kingdom of God, where corruption and death are no more.

If human beings were to treat one another's personal property the way that they treat their environment, we would view that behavior as anti-social. We would impose judicial measures necessary to restore wrongly appropriated personal possessions. It is, therefore, appropriate for us to seek ethical, and even legal recourse where possible, in matters of ecological crimes.

ENVIRONMENTAL SIN

It follows that to commit a crime against the natural world is a sin. For human beings to cause species to become extinct and to destroy the biological diversity of God's creation; for human beings to degrade the integrity of the earth by causing changes in its climate, by stripping the earth of its natural forests, or destroying its wetlands; for human beings to injure other human beings with disease; for human beings to contaminate the earth's waters, its land, its air, and its life, with poisonous substances—these are sins.[19]

In prayer, we ask for the forgiveness of sins committed both willingly and unwillingly. And it is certainly God's forgiveness, which we must ask, for causing harm to His own creation. In this way, we begin the process of healing our worldly environment, which was blessed with beauty and created by God. Then, we may also begin to participate responsibly, as persons making informed choices in both the integrated whole of creation, and within our own souls.

In just a few weeks the world's leaders will gather in Kyoto, Japan, to determine what, if anything, the nations of the world will commit to do in order to halt climate change. There has been much debate back and forth about who should and should not have to change the way they use the resources of the earth. Many nations are reluctant to act unilaterally. This self-centered behavior is a symptom of our alienation from one another and from the context of our common existence.

---

19. Patriarch Bartholomew was the first religious leader and thinker to dare broaden the concept of sin—beyond individual and social implications—to include environmental abuse.

ENVIRONMENTAL ETHICS

We are urging for a different and, we believe, a more satisfactory ecological ethic. This ethic is shared with many of the religious traditions represented here. All of us hold the earth to be the creation of God, where He placed the newly created human "in the Garden of Eden to cultivate it and to guard it" (Gen. 2.15). He imposed on humanity a stewardship role in relationship to the earth. How we treat the earth and all of creation defines the relationship that each of us has with God. It is also a barometer of how we view one another. For, if we truly value a person, we are careful as to our behavior toward that person. Yet, the dominion that God has given humankind over the earth does not extend to human relationships. As the Lord said, "You know that the rulers of the nations lord it over them, and their great ones are tyrants over them. It will not be so among you; but whoever wishes to be great among you must be your servant, and whoever wishes to be first among you must be your slave; just as the Son of Man came not to be served but to serve, and to give his life as a ransom for many" (Mt. 20.25–28).

It is with that understanding that we call on the world's leaders to take action to halt the destructive changes to the global climate that are being caused by human activity. And we call on all of you here today to join us in this cause. This can be our important contribution to the great debate about climate change. We must be spokespeople for an ecological ethic that reminds the world that it is not ours to use for our own convenience. It is God's gift of love to us and we must return his love by protecting it and all that is in it.

Keynote Address at the International Ecological Symposium,
Kathmandu, Nepal, November 2000

SACRED GIFTS TO A LIVING PLANET

The purely religious and non-administrative character of the ministry that we exercise also determines the framework of our potential intervention and contribution for the protection of the natural environment. We are unable to assume necessary measures. Nor can we instigate forceful reactions against those who negatively influence the environment. We are simply in a position to address free and conscientious people in order to suggest to them what is correct according to our faith so that they may be persuaded to conform freely and willingly, as their obligation and responsibility dictate in regard to this matter.

Therefore, what we shall say here follows first from our human perspective and consequently from our co-responsibility for the future of our planet and for human life on this planet. So our words should be acceptable to all people of good will, irrespective of their religious conviction, because they are surely, in our opinion, the proper conclusions of rational thought. Furthermore, what we shall say also flows from the tenets of our Christian faith and from our worldview according to the perspective of this faith, that is to say according to the way in which we regard the world, humanity, creation, and history. We recognize that there exist numerous and diverse theoretical beliefs and philosophical opinions concerning this worldview. However, we believe that the foundation of our practical conclusions, reflected in our personal and collective behavior, is able to assist all people of good will to conform their behavior to the suggested ethical model. This should be the case even if the religious origin of the suggested ethic does not coincide with certain people's opinions about particular details.

We are obliged to clarify the fact that, for us the demands of our faith are of primary value. Yet, we begin with the more general human demands because we are also addressing distinguished delegates who do not share our religious conviction, and for whom the weight of any argument does not lie in its religious connection but in its rational cohesion.

FUNDAMENTAL DATA

Before exposing our arguments on this matter, we consider it necessary to focus your attention on certain data, well known of course to everyone but nevertheless still worthy of particular emphasis.

(i) *The planet earth, on which all of us necessarily live, constitutes a minute part of the universe.* No matter how great its dimensions appear to us by comparison with human dimensions, it is truly very small when compared to the rest of the universe.

(ii) *This planet is not separated by natural boundaries into close-knit compartments. The legal borders of nations are unknown to nature.* The winds that blow and transfer particles and gases from place to place do not seek anyone's commission in order to follow the direction of the natural laws. The same also applies to the waters, to temperatures, and to every form of radiation.

(iii) *The range of every occurrence is very wide, namely worldwide, and in a sense universal. This means that if one spot of pollution is emitted in one point of the earth, its effect is felt throughout the world.* Of course, the effect is not immediately perceived in the case of a small trace of pollution. It becomes more apparent in the case of extended pollution. Yet the effect is the same. For example, scientists expect that in the coming decades the average temperature on the surface of the earth will increase by several degrees. This fact will also result in the melting of ice, the raising of the sea level, greater rainfall and floods in colder regions, and more dryness and desert in warmer regions. Such effects are not the result of actions occurring in the affected areas alone, but in any and every part of the world. For, as a result of the natural law of Aristotelian entelechy[20] and entropy,[21] of the equation of differences, warm temperatures in any part of the world, and derived from any source at all, are added to temperatures resulting from other sources. Accordingly, these temperatures are conveyed throughout the surface of the earth. What is of importance is not so much that, according to specialists, it takes up to two years for air pollution in a particular place to spread throughout the atmosphere of the globe. What is of importance rather is that the laws of chaos imply that any unhealthy intervention in one region of the atmosphere is similarly shared throughout the world.

(iv) *The consequences of a polluting action, after a necessary period, equally affect every person throughout the world, including the responsible perpetrator, as well as a boundless number of innocent victims.* So it is inconceivable for the perpetrator to protect himself from the consequences of his action, and it is impossible to know who will ultimately be the victims of such action. However it is fact that humanity is collectively and in its entirety damaged. Certainly, the density of pollution, and therefore the risk of damage, is greater in the region and at the time closer to the emission. Yet, whatever remains after the natural function of self-purging, which

---

20. Sometimes translated as "actuality," it is used by Aristotle to contrast with potentiality (or *dynamis*) and is a principle by which he defines motion or causality.

21. A thermodynamic property and measure of energy.

occurs through various means and natural functions, is defused and divided into the entire planet.

(v) Already from the period of the first appearance of the individualistic Roman Law, the owner of any property possessed the right to enact whatever he so wished on this property irrespective of whether this disturbed or damaged his neighbor. Yet, early on, certain cracks began to appear in this absolutely individualistic principle. One example of this may be found in the foresight to forbid certain dangerous emissions. Inasmuch as certain social obligations and connections were also gradually related to the ownership of property, the disturbing use of land increased, and, as a result, many places have already established regulations of good neighborhood which are socially acceptable and desirable. *This means that societies recognize that the individualistic principles formerly pervading civil law are today becoming more social.* Therefore, the individualistic principles that in many ways govern international law are also able to become more social through our influence as members of societies.

We have already noted that our planet is so small and so bereft of close-knit compartments that every material and human act that is damaging for the environment bears consequences of worldwide extent. And from our observations about the development of law and legal institutions, our desire is to emphasize accordingly the worldwide effect of every change in the spiritual attitude and conduct of any one citizen in regard to the environment.

The necessary conclusion is that any non-governmental effort to change the attitude of citizens, even if it appears to have only limited efficacy, has profound significance for the environment and its improvement.

A MUTUAL RELATIONSHIP WITH NATURE

Let us now turn to the environmental imperatives that derive from our human nature. Herodotus[22] mentions that in some region of the classical world, the people regarded it as blasphemy, as contradicting the very will of the gods, to pollute the rivers. These people, according to some modern thinkers, might be considered culturally underdeveloped in as much as

---

22. Herodotus (484–425 BC), the "father of history."

they did not experience the development of our technical civilization. Yet their spiritual sensitivity and refinement, when compared with the corresponding sensitivity of contemporary and civilized human beings that pollute the rivers with tons of poisonous substances as they sail through them, must be considered exceptional and excellent, whereas we would surely fail by their standards. The possibility that the religious basis of their behavior may today not be accepted as it was then described or even believed, does not undermine the character of their behavior that was socially perfect and especially commendable in this regard.

It has also been observed with great conviction that the majority of deserts, and especially those in Mesopotamia and other formerly inhabited regions, are the result of human actions, such as deforestation for the sake of cultivation, fires caused by arson, the desalination of the earth and the abuse of nature. It is well known that acid rain, which comes from sulfur dioxide produced in a mining station of Sudbury in Ontario, Canada, has from 1888 to this day destroyed two million acres of surrounding coniferous forests. The nearby region does not have even a trace of vegetation.

Nevertheless, it is not only the nearby regions that are affected by any source of pollution. We know that air pollution produced in England affects the biotic communities of Sweden; pollution in the Great Lakes affects the residents of Canada and the United States; and the radioactive Strodium 90 has been traced in bodies of the distant Eskimo people to a far greater extent than in populations living much closer to the points of emission. This is because the radioactive element was absorbed into the lichens consumed by the caribou that constitute the primary source of food for the Eskimos. Furthermore, we are all aware of the transportation of radio energy from the accident of the nuclear power station in Chernobyl and of the statistical predictions by scientists about the increase of cancer even in populations of countries at a significant distance from this source.

These are but a few of the many possible indications that any change in the environment is a matter concerning all people and all regions in the world. And so all of us ought to become conscious of our collective obligation to conform to everything demanded for the sake of the protection of the environment. This obligation is fundamentally twofold. First, we ought actively to avoid destroying or polluting the environment, and actively endeavor to restore it and improve it. And second, we ought passively to reject the use of products whose production burdens our environment, or at least to use these products rationally, if this is absolutely necessary. We could add a third obligation, namely the effort to render all people aware of these responsibilities. For, although the potential influence of each individual may appear to be limited, the collective influence of all people together is limitless. This environmental ethic is imperative of rational thinking and must be understood as equivalent to the obligation of self-preservation.

A MODEL OF BEHAVIOR

Now we come to what the Orthodox Christian Church believes and teaches. Let us begin by mentioning a characteristic commandment of the founder of the Christian Church, of our Lord Jesus Christ. Multitudes followed the Lord into the desert in order to hear His teaching and receive healing for their illnesses. Christ blessed five loaves of bread and two fish, instructing His disciples to share these among the five thousand men, gathered with their wives and children. All of them ate, as we are told, and were filled. Up to this point, the narrative describes a miracle. Yet in continuation, the miracle-worker said to His disciples: "Gather up the fragments that remain, so that nothing is lost" (Jn 6.12). The commandment to gather up the remainders "so that nothing is lost," especially as it comes from the mouth of their Creator, constitutes a model of behavior that is most useful for our time when the refuse of certain large cities, rejected as trash, could suffice to nourish entire populations.

You are undoubtedly familiar with the charitable commandment of love and mercy, which is taught by the Orthodox Church. You are perhaps even aware that the Ecumenical Patriarchate, as the first throne among the worldwide Orthodox Churches, has, at the dawn of the third millennium from the Birth of Christ, placed at the center of its attention the urgent problem of our times, namely the preservation of balance in the natural environment of our planet. We are absolutely convinced that an effective approach to this problem requires not only the intervention of governments, but also the cultivation of an ethic based on an understanding of the relationship between humanity and nature, which, beyond this ethic, derives from the coexistence of all human beings.

For us, this ethic stems from our faith in God and, as we believe, from the event of creation. We accept that God created the universe out of nothing and out of love. As the crown of this creation, God formed Adam, whom He established in the Garden of Eden, the earthly paradise. The delight in the goods of Paradise was not an end to itself. Adam's pleasure in paradise was not due to the enjoyment of the material goods therein; it was not an animal satisfaction of instincts. God fashioned Adam into a personal and spiritual being, created out of two elements. One element was drawn from the material creation made by Him, "dust from the dust" of the earth, and therefore constituted the human body. The other element was a created spirit, similar to the uncreated spiritual essence of God, and so Adam was created "in the image and likeness" of God, namely endowed with all the good attributes of the persons of the Triune God. These attributes include personal being, mind, freedom, love, judgment, will, and so forth. And the delight of humanity in Paradise was founded on a personal relationship with the Creator, a relationship characterized as love and incorporating full trust in God.

THE COMMANDMENT TO CARE

While being a spirit in image of the Spirit of God, the human person was not created perfectly divine. Rather, the human person was created with the possibility of becoming like God through gradual progress derived from personal and voluntary asceticism, as well as through the intervention of divine grace, namely the uncreated energy of God. This ascetic discipline embraced three things: first, work in Paradise; second, the keeping of Paradise; and, third, the fulfillment of a commandment to avoid consuming one fruit. In reference to these three points, the basis of the desired behavior was love, and the desired goal from God was the preservation and increase of humanity in personal love. This communion of humanity with God would render us partakers of divine nature and blessedness, namely divine by grace although not in essence.

Insofar as the human person also participates in the created, material nature by means of the body, he or she would be called to assume the material creation within which it has been established as crown. And so, by assuming material creation as nurture and harmony, the human person would raise it up in thanksgiving to God, from whom everything was received. Nature was planned by God to offer of its own, and especially through its cultivation and protection by humanity, everything that is necessary for the preservation and pleasure of humanity. However, the relationship of humanity toward nature was not one of possession. Rather, it was a relationship of a person toward a gift, and especially toward God, the giver of all that is good. Therefore, humanity must remember at all times that it holds this gift only within certain boundaries established by the giver, and that within these boundaries there are two conditions.

The first condition is the requirement to protect the gift, namely to preserve nature harmless, and to consume only its fruit. The second condition is the requirement not willfully to consume every fruit, but to be self-restrained and to abstain from certain fruits. Both the protection and the self-restraint, which in ecclesiastical terminology is called *ascesis*, were not imposed as authoritative commandments with appropriate consequence in the case of their transgression. Instead, they were offered as suggestions of love that ought to be preserved out of love. In this way, love is preserved alive, as a personal relationship and mutual communion between God and humanity. For the true nature of God is love; and the original nature of the human person was also endowed with love, seeking to be established in love so as to become divine by grace. Therefore, that which would most liken humanity to God was precisely the establishment of humanity in love in the same way as God Himself is stable and unchanging in this love.

The first-created human being, however, misused the God-given freedom, preferring alienation from God and attachment to God's gift. Consequently, the double relationship of humanity to God and creation was canceled in regard to its direction toward God, leaving humanity preoccupied with creation alone. In this way, the human blessedness derived from the love between God and humanity ceased, and humanity sought to fill this void by drawing from creation the blessedness that was lacking. From thankful user, the human person became greedy abuser. Humanity sought from creation to offer that which it could not, namely the blessedness that was missing. The more that humanity feels dissatisfied, the more it also demands from nature. Yet the more it demands of nature to offer, the more humanity recognizes that the goal has escaped its grasp. The soul's emptiness cannot be filled with the world's possessiveness. Nor again can it be achieved by the acquisition of material goods, for it does not result from lacking these goods. The soul's emptiness results from a lack of love and familiarity with God. The lost paradise is not a matter of lack in created goods, but of deprivation of love toward the Creator.

Christ came into the world in order to restore, and He did restore the possibility of our love toward God. For, as fully human (and divine), He fully loved God (the Father), becoming an example of the loving relationship between humanity and divinity. Those who sincerely and correctly believe in Him and love God practice keeping the original commandments of God. They practice the commandments to work, to keep the natural creation from any harm, and to use only its fruits, indeed those fruits that are absolutely necessary to use, taking proper care "that nothing is lost," and becoming conscientious models of environmental care.

An Orthodox Christian ethic, therefore, emanates from the worldview of humanity, creation, and God that we have very briefly presented. All other Christian exhortations about the proper way of life stem from the conscious effort of human beings to cease hoping in creation and turn their hope to the Creator of all. When this attitude is adopted, humanity will be satisfied with much fewer material goods and will respond with greater sensitivity to the nature that nurtures us. Humanity will then be concerned about loving all people, and will not seek to satisfy individualistic and egotistic ambitions.

The Orthodox Church penetrates the reason of every being, namely the origin and purpose of every created thing, discerning the complete plan of God from the first moment of creation to the end of the world, considering this as an expression of absolute love and offering. The world was created "very good" (Gen. 1.31) in order to serve the mind of God and the life of humanity. However, it does not replace God; it cannot be worshipped in the place of God; it cannot offer more than God appointed it to offer. The Orthodox Church prays that God may bless this creation in order to offer seasonable weather and an abundance of fruits from the earth. It prays that God may free the earth from earthquakes, floods, fires, and every other harm. In recent times, it has also offered supplications to God for the protection of the world from destruction caused by humanity itself, such as pollution, war, overexploitation, exhaustion of waters, changes in environmental conditions, devastation, and stagnation.

MORE THAN MERE PRAYER

The Ecumenical Patriarchate does not, however, rely only on supplication to God to improve the situation. Starting from God, as it is always proper to seek His blessing, the Ecumenical Patriarchate works intensely in every possible way to alert everyone to the fact that the greed of our generation constitutes a sin. This greed leads to the deprivation of our children's generation, in spite of our desire to bequeath to them a better future.

The relative environmental activity of the Ecumenical Patriarchate has been revealed, among other ways, in the convocation also of international ecological Seminars in the historical Holy Monastery on the island of Halki, which began a systematic theological study of the ecological crisis. There has also been participation on our part in numerous conferences convened to discuss these matters. Our environmental initiative is further reflected in the organization of three floating symposia attended by scientists from all over the world. In admirable harmony of spirit, we advanced from the study of general themes to the examination of particular problems.

In September of 1995, together with His Royal Highness Prince Philip, Duke of Edinburgh, we organized a floating symposium in the Aegean, starting from the island of Patmos, in order to celebrate the 900th anniversary of the Book of St. John's Revelation. During this symposium, faith and understanding, religion and science, spirit and word approached, from different perspectives, one and the same purpose, namely the protection of the environment. The success of this symposium inspired and encouraged us to organize a second international floating symposium in the area of the Black Sea, in light of the ecological destruction of that region.

The deliberations of the Black Sea Symposium made it abundantly clear that the pollution of this sea largely depends on the pollutants washed up by the rivers. The warm hospitality and fervent reception expressed by the inhabitants of this region toward the delegates of our symposium was deeply moving. Having studied the existing problems, we decided to continue our research by organizing yet another symposium, in the context of which we traveled by ship along the Danubian countries.

Beyond the study of pollution and the search for a solution to the dangerous conditions and constructions, and the pain and the poverty all along this great river, a further purpose of this symposium was the healing of this region of Europe that has been plagued with terrible ordeals. In the context of our limited capacity, we proposed to define the principals of free communication, of mutual respect, and of peaceful coexistence among the people of this region.

In the aftermath of the symposium, it is important for us to search for its practical results. These results give rise to hope for an activation of interest in the environment. At the conclusion of the Black Sea Symposium, a systematic environmental education created a network of concerned clergy, journalists, and teachers from the Black Sea region. Today, an initiative to create a similar environmental network is gaining ground in the Danubian countries, with the participation of various Churches, in conjunction with the Danube Carpathian Program of the Worldwide Wildlife Fund (WWF). This network will preserve alive the ecological initiative and strengthen the cooperation among the peoples of the region.

A further symposium is scheduled for the Baltic Sea. This symposium will direct its attention to one of the more burdened coastal environments of our planet. The Baltic Sea has suffered both from the development and wealth of some countries in the region, as well as from the poverty of certain others. This symposium will also endeavor to underline the significance of a common effort and formation of a common ethic. It will remind the world that the sacred creation is not our property, and so we do not have the right to use it according to our desires. It is the gift of God's love to us, and we are obliged to return his love by protecting everything embraced by this gift.

We would like to conclude by expressing our prayer and hope for the future of humanity. We fervently pray that peace and harmony will prevail, not simply as an absence of conflicts or as a temporary truce in confrontations, but as a stable condition for the future of our planet. To this purpose, however, all of us are required to work, and to work together, irrespective of religious convictions. Therefore, we fraternally call upon all religious leaders to adopt the effort for the protection of the natural environment and to inspire their faithful in the religious and humanitarian obligation to participate in this endeavor, both actively and passively.

Address to the Scenic Hudson, New York, November 13, 2000

AN AWARD FOR THE PAST

Scenic Hudson, well known for its efforts to protect Storm King Mountain, was one of the first voices of environmentalism in this land. Thirty years later, it remains one of the most efficient organizations of its kind. We wish you many years of fruitful service in this critical and vital issue for all humankind.

Similarly, for many years now, the Orthodox Ecumenical Throne has devoted itself to the service of the protection of the environment. With great interest and sincere anxiety, we have followed the efforts to address the destructive side effects of humanity upon the world of nature. These effects have also a negative influence upon human beings themselves. With much trepidation, we now realize the dangerous consequences of human apathy concerning the survival of creation, which include the survival of humankind itself.

It is for this reason that we are accepting this award in the name of my illustrious predecessor, the late Ecumenical Patriarch Dimitrios. He is the one who invited the entire world to offer, together with the Holy Great Church of Christ, the Ecumenical Patriarchate, prayers of thanksgiving, but also prayers of supplication for the protection of God's gift of creation. Thus, as of 1989, the beginning of the new ecclesiastical year, commemorated and celebrated September 1st, which has been designated for all Orthodox Christians throughout the world as a day of prayer for the protection of the environment.

THE CONTRIBUTION OF ORTHODOXY

Beyond any stereotype, the following question would be absolutely justified: In what way can Orthodoxy contribute to the movement for the protection of the environment? By the grace of God, there is one concrete response. We believe that through our *unique liturgical and ascetical ethos*, the spiritual teaching of the Orthodox Church may provide important theoretical and deontological direction for the care of our planet earth.

The spiritual root of our pollution, and our sin against the world, consists in our refusal to face life and the world as God's gift to humankind, which humans have to utilize with discernment, with respect, and with thanksgiving. In the Orthodox Church, we call this the "Mystery (or, the Sacrament) of the Holy Eucharist," in which we return in thanksgiving to Christ the entire creation. We do this out of gratitude to Him, who was crucified on behalf of the world; we do this in the framework of the eucharistic celebration, during which His sacrifice on the Cross is repeated in a sacramental way. "We offer Thine own [gifts] from Thine own,"[23] we exclaim as we offer the bread and wine, basic elements of the natural creation, in order to be changed by the Holy Spirit and become the Body and Blood of Christ, namely the gifts of our continuous communion with God.

Thus, we believe that our first duty is to stimulate the human conscience to realize that, when human beings utilize the resources and the elements of our planet, they should do this in a devout and eucharistic way. Ultimately, to the benefit of our posterity, we should consider every act through which we abuse the world as having an immediate negative effect upon the future of our environment, as relating to the prosperity of our world. The heart of the relationship between humans and their prosperity is found in the relationship between humans and their environment. The way in which we face our environment reflects upon the way we behave toward one another; more specifically, it reflects upon the way in which we relate to our children, to those born and those who are yet to be born.

A SEAMLESS GARMENT

Human beings and the environment compose a seamless garment of existence, a multi-colored cloth, which we believe to be woven in its entirety by God. Human beings are created by God as spiritual beings, reflecting the image of God (Gen. 1.26). However, human beings are also created by God from material nature, from the dust of the earth. Consequently, we are called to recognize this interdependence between our environment and ourselves. This interconnectedness between our environment and ourselves lies in the center of our liturgy. In the seventh century, St. Maximus the Confessor described this liturgy as being beyond a divine or mere human liturgy. We cannot avoid our responsibility toward our

---

23. From the prayer of the *anaphora* in the Divine Liturgy.

environment and toward our fellow human beings, who are negatively affected as by its deterioration.

In the Orthodox Church, there is also the ascetical element, which requires voluntary restraint regarding the use of material goods, and which leads to a harmonious symbiosis with the environment. We are required to practice "restraint" (the theological term in Greek is *enkrateia*). When we curb our own desire to consume, we guarantee the existence of treasured things for those who come after us, and the balanced functioning of the ecosystem. Restraint frees us from selfish demands so that we may offer and share what remains, placing it at the disposal of others. Otherwise, we are characterized by avarice, which has its roots in lack of faith and the worship of matter, which we consider idolatry. Restraint is an act of humility and self-control, of faith and confidence in God. It is also an act of love. There are Christians who voluntarily deprive themselves of their due portion and exercise restraint in order to share with those who have a greater need. This ascetical spirit gives us the example according to which we may live by being satisfied with what is needed, without collecting needless things, without consumerism that leads to exploiting and lording it over nature. This voluntary ascetical life is not required only of the anchorite monks. It is also required of all Orthodox Christians, according to the measure of balance. Asceticism, even the monastic such, is not negation but a reasonable and tempered utilization of the world.

THE IMPACT OF AN ORTHODOX MONK

It was correctly stated that a human being who has lost the self-consciousness of the divine origin, of humanity created in the likeness of God, has also lost the sense of divine destiny. In a word, such a person has lost self-esteem as a human being reflecting the image of God and has tried to make up for this loss by increasing material goods, over which one may have control. *When one's "being" is decreased, one's need for "having" is increased.* Consequently, the consciousness that a Christian has of his or her existence renders superfluous the need for consumerism and the accumulation of material goods. For this reason the seventh-century hermit on Mt. Sinai, St John of the Ladder, said: "A monk without possessions is master of the entire world." And St. Paul recommends the avoidance of avarice when he writes: "As we have food and clothing, let them suffice to us" (1 Tim. 6.8).

Therefore, the Orthodox ascetical life is not an escape from society and the world, but a way of self-sufficient social life and behavior, which leads to a reasonable use and not an abuse of material goods. The opposite worldview leads to consumerism, to an excessive drawing from the productive ecosystem, a reversal of its balance, its destruction, and, in the long run, to an inability to survive as environment and as race. The Orthodox ascetic attitude seems to be passive; it appears not to impose any method of dealing with and solving the environmental problems of our time. However, just as the individual actions of tens of thousands of members of society produce great pollution, so the voluntary restraint of an Orthodox monk is of great benefit for all.

Repentance over our past mistakes regarding the environment is indispensable and useful. Unfortunately, humanity has become intoxicated by its technological possibilities and behaves tyrannically toward the environment. Humankind ignores the fact that silent nature will take its revenge—perhaps slowly, almost unnoticeably, but inevitably and surely.

ENVIRONMENTAL RIGHTS

Unfortunately, avarice and excessive exploitation, with no regard for their consequences, are a usual phenomenon. If we were to behave toward the possessions of our fellow human beings in the same way that we behave toward the environment, we would suffer legal sanctions and expect compensation for damages. We would have to use legal remedies to restore the damages and return the stolen property to its legitimate proprietor. Our behavior would be characterized as anti-social if we were to offend our fellow human beings.

From the declarations of human rights, we gather on the one hand that most of the environmental goods—such as air, water, and the like—are not able to become private property; on the other hand, their possession demands a proportionate legal and canonical obligation, which cannot be ignored. Whatever use by the possessor contradicts the social obligation and usefulness of this good is prohibited as abusive and is subject, or should be subject, to legal sanctions.

The imposition of such sanctions does not belong to the realm of the Church, which addresses itself to the self-consciousness of humanity and requires voluntary compliance. This compliance is accompanied by the obvious alarm that disrespect toward nature constitutes a sin against the love for God and humanity as well as against God's creation. According to Scripture, "the wages of sin are death" (Rom. 6.23). At this point, this is confirmed from our everyday experience of the chain reactions of the environmental destruction: changes in the climate, the stripping of the earth from its forests, torrential rainfalls, floods, mudslides; the consequence of all these is death. Atomic explosions, radioactivity, cancerous births: Again, the consequence of these is death. Toxic wastes, pollution of the air, water and the ground, introduction of toxic substances into the cycle of life: Once more, the consequence is death. Dispersion into the atmosphere of gases that damage the ozone, augmented infrared radiation that damages human health; this, too, leads to death.

For all these causes of death, which are the direct result of our own doing, and of which we are not conscious that we are the cause, in our prayers we ask for God's forgiveness. Our responsibility for whatever happens around us is an unavoidable given. We not only destroying the beauty of created nature, but we are also bringing harm and death to our fellow human beings. To remedy the situation, we should become conscious of this great sin; we should allow it to become an important motivation to ameliorate our environmental behavior. Then, the goal of our common ecological responsibility will also become increasingly socially acceptable. Then, perhaps we will begin responsibly to participate as individuals with conscious choices, whether in the context of the environment or in the context of our souls.

ECUMENICAL IMPERATIVE

For all these reasons, we address ourselves to the leaders of the world and pray that they take the necessary measures so that the catastrophic changes of climate, caused by human activity, may be reversed. We should propagate an ecological ethic, which should remind us that the world is not ours for us to use as we please. It is a gift of God's love to us. It is our obligation to return that love by protecting it with whatever responsibilities this may entail.

This common purpose unites all human beings, in the same way as all the waters of the world are united. In order to save a sea, we must save all the rivers and oceans. God created heaven and earth as a harmonious totality; consequently, we also have to face creation as harmonious and interdependent whole. For us at the Ecumenical Patriarchate, the term "ecumenical" is more than a name: It is a worldview, and a way of life. The Lord intervenes and fills His creation with His divine presence in a continuous bond. Let us work together so that we may renew the harmony between heaven and earth, so that we may transform every detail and every element of life. Let us love one another. With love, let us share with others everything we know and especially that which is useful in order to educate godly persons so that they may sanctify God's creation for the glory of His holy name.

As a symbol to remind us of this responsibility—that each of us must do our part so that we may keep our natural environment as it has been handed down to us by God—we present you with this parchment containing the inscription from Holy Scripture of God's commandment to the first-created people placed in the Garden of Eden. It is the commandment to "work and keep the earth" (Gen. 2.15) of the garden. This is also the content of our message, addressed to every human being. Let everyone work to produce material goods from nature; but also, let everyone keep its integrity and keep harmless, as God commanded human beings to do.

Address at the Brookings Institution, Washington, D.C., November 4, 2009

SAVING THE SOUL OF THE PLANET

It is a pleasure and a privilege to address members and guests of this renowned center of political study and thought. At first glance, it may appear strange for the leader of a religious institution concerned with *spiritual values* to speak about the environment at a secular institution that deals with *public policy*. What exactly does preserving the planet or promoting democracy have to do with saving the soul or helping the poor? It is commonly assumed that ecological issues—global climate change and the exploitation of nature's resources—are matters that concern politicians, scientists, technocrats, and interest groups.

The Ecumenical Patriarchate is certainly no worldly institution. It wields no political authority; it leads by example and by persuasion. And so the preoccupation of the Orthodox Christian Church and, in particular, her highest spiritual authority, the Ecumenical Patriarchate, with the environmental crisis will probably come to many people as a surprise. But it is neither surprising nor unnatural within the context of Orthodox Christian spirituality.

Indeed, it is now exactly twenty years since our revered predecessor, Ecumenical Patriarch Dimitrios, sparked the ecological initiatives of our Church by issuing the first encyclical encouraging our faithful throughout the world to pray for and preserve the natural environment. His exhortation was subsequently heeded by the member churches of the World Council of Churches.

What, then, does preserving the planet have to do with saving the soul? Let us begin to sketch an answer by quoting an Orthodox Christian literary giant, Fyodor Dostoevsky, echoing the profound mysticism of Isaac the Syrian in the seventh century through Staretz Zossima in *The Brothers Karamazov*:

> Love all God's creation, the whole of it and every grain of sand. Love every leaf, every ray of God's light! If you love everything, you will perceive the divine mystery in things. Everything is like an ocean, I tell you, flowing and coming into contact with everything else: touch it in one place and it reverberates at the other end of the world.[24]

This passage illustrates why, with respect to the priority and urgency of environmental issues, we do not perceive any sharp line of distinction between the pulpit and this lectern. One of our greatest goals has always been to weave together the seemingly disparate threads of issues related to human life with those related to the natural environment and climate change. For as we read the mystical teachings of the Eastern Church, these form a single fabric, a seamless garment that connects every aspect and detail of this created world to the Creator God that we worship.

THE ENVIRONMENT CONNECTS US

For how can we possibly separate the *intellectual* goals of this institution—namely, the advancement of democracy, the promotion of social

---

24. *The Brothers Karamozov* (Harmondsworth, UK: Penguin, 1982), vol. 1, 375–376.

welfare, and the security of international cooperation—from the *inspirational* purpose of the church to pray, as we do in every Orthodox service, "for the peace of the whole world," "for favorable weather, an abundance of the fruit of the earth," and "for the safety of all those who suffer"?

Over the past two decades of our ministry, we have come to appreciate that one of the most valuable lesson to be gained from the ecological crisis is neither the political implications nor the personal consequences. Rather, this crisis reminds us of the connections that we seem to have forgotten between previously unrelated areas of life.

It is a kind of miracle, really, and you do not have to be a believer to acknowledge that. For, the environment unites us in ways that transcend religious and philosophical differences as well as political and cultural differences. Paradoxically, the more we harm the environment, the more the environment proves that we are all connected.

Our dear friend Strobe Talbott in New York last week, during the 9th Annual Orthodox Christian Prayer Service sponsored by the Eastern Orthodox and Ancient Oriental Churches for the United Nations Community, you spoke of the need to "redefine the international [or global] community . . . beyond the horizons of our life span."

The global connections that we must inevitably recognize between previously unrelated areas of life include the need to discern connections between the *faith communities*. We must also perceive the connections between all *diverse disciplines*; climate change can only be overcome when scientists and activists cooperate for a common cause. And, finally, we can no longer ignore the connections in our hearts between *the political and the personal*; the survival of our planet depends largely on how we translate traditional faith into personal values and, by extension, into political action.

That is why the Orthodox Church has been a prime mover in a series of inter-disciplinary and interfaith ecological symposia held on the Adriatic, Aegean, Baltic, and Black Seas; along the Amazon and Danube Rivers; as well as on the Arctic Ocean. The last of these symposia concluded only a few days ago in New Orleans, seeking ways to restore the balance of the great Mississippi River.

THE CONSEQUENCES OF OUR ACTIONS

The mention of New Orleans brings to mind another truth. Not only are we all connected in a seamless web of existence on this third planet from the Sun, but there are profound analogies between the way we treat the earth's natural resources and the attitude we have toward the disadvantaged. Sadly, our willingness to exploit the one reflects our willingness to exploit the other. There cannot be distinct ways of looking at the environment, the poor, and God.

This is one of the reasons why we selected New Orleans as the site of our latest symposium, and this is why our visit there was, in fact, the second since the devastation of Hurricane Katrina. There, images of poverty abound, too close for comfort. We witnessed them in August of 2005 on the Gulf of Mexico; they are still evident over four years later—not only sealed forever in our memory, but soiling Ward 9 to this day! How could the most powerful nation on earth appear so powerless in the face of such catastrophe? Certainly not because of lack of resources! Perhaps because of what St. Seraphim of Sarov[25] once called "lack of firm resolve."

The truth is that we tend—somewhat conveniently—to forget situations of poverty and suffering. And yet, we must learn to open up our worldview; we must no longer remain trapped within our limited, restricted point of view; we must be susceptible to a fuller, global vision. Tragically, we appear to be caught up in selfish lifestyles that repeatedly ignore the constraints of nature, which are neither deniable nor negotiable. In New York, Secretary Talbott referred to the need for "the singular 'I' to reflect the plural 'we'—[that is to say] the ecumenical or global reality." We must relearn the sense of connectedness. For we will ultimately be judged by the tenderness with which we respond to human beings and to nature.

Surely one area of common ground, where all people of good will—of all political persuasion and every social background—can agree is the need to respond to those who suffer. Even if we cannot—or refuse to—agree on the root causes and human impact on environmental degradation; even if we cannot—or refuse to—agree about what would define success in sustainable development, no one would doubt that the consequences of climate change on the poor and disadvantaged are unacceptable. Such denial would be inhumane at the very least and politically disadvantageous at worst.

---

25. St. Seraphim (1759–1833) is a beloved mystic and staretz of the Orthodox Church.

Of course, poverty is not merely a local phenomenon; it is also a global reality. It applies to the situation that has existed for so long in such countries as China, India, and Brazil? To put it simply, someone in the "third-world" is the most impacted person on the planet, yet that person's responsibility is incomparably minute: What that person does for mere survival neither parallels nor rivals our actions in the "first-world."

AN INCONVENIENT TRUTH

Many argue that the wealthy nations of the West became so by exploiting the environment—they polluted rivers and oceans, razed forests, destroyed habitats, and poisoned the atmosphere. But now that that the poorer nations are developing and improving the quality of life for their citizens—as the West did during the nineteenth and twentieth centuries—all of a sudden the rules are being changed and developing nations are being asked to make sacrifices the nations of the West never made as they were developing. They are being asked to reduce their impact on the environment—in other words, to curb their development. They are being asked to drive fewer cars, consume less oil, build fewer factories, raze fewer forests, and harm fewer habitats—all in the name of protecting the environment.

Brothers and sisters—this simply cannot be. Not only is it unfair to ask the developing nations to sacrifice when the West does not—it is futile. They care not what we say—they watch what we do. And if we are unwilling to make sacrifices, we have no moral authority to ask others, who have not tasted the fruits of development and wealth, to make sacrifices.

Fortunately, the West, and in particular America, is now showing that it recognizes this "inconvenient truth"[26]—that if we are to save our planet, sacrifices must be made by all. The Obama administration, as you know, has been very active in this regard. The President has signed an Executive Order challenging government agencies to set 2020 greenhouse

---

26. Title of a documentary and book by, among others, former U.S. Vice President Al Gore, subtitled *The Planetary Emergency of Global Warming and What We Can Do About It* (Emmaus PA: Rodale Press, 2006).

reduction goals, and using the government's $500 billion per year in purchasing power to encourage development of energy-efficient products and services.

There are also many promising developments at the global level. Representatives of the 16 countries that emit the highest levels of greenhouse gases met recently in London to discuss the amount of aid they will give less-developed nations to help them adopt cleaner energy technology. And there are growing expectations that meaningful progress can be made as a result of the United Nations Climate Change Conference scheduled to take place in Copenhagen next month.

Sacrifices will have to be made by all. Unfortunately, people normally perceive "sacrifice" as loss or surrender. Yet, the root meaning of the word has less to do with "going without" and more to do with "making sacred."[27] Just as pollution has profound spiritual connotations related to the destruction of creation when disconnected from its Creator, so, too, sacrifice is the necessary corrective for reducing the world to a commodity to be exploited by our selfish appetites. When we sacrifice, we render the world sacred, recognizing it as a gift from above to be shared with all humanity—if not equally, then at least justly. Sacrifice is ultimately an expression of gratitude (for what we enjoy) and humility (for what we must share).

THE CENTER FOR ENVIRONMENT AND PEACE

For our part, in addition to our international ecological symposia, the Orthodox Church has decided to establish a center for environment and peace. Hitherto, the Ecumenical Patriarchate has endeavored to raise regional and global *awareness* on the urgency of preserving the natural environment and promoting inter-religious dialogue and understanding. Henceforth, the emphasis will be *educational*—on the regional and international levels.

The *Center for Environment and Peace* will be housed in a historical orphanage, on Prinkipos (Büyükada), one of the Princes' Islands near Istanbul. The building was once the largest and most beautiful wooden edifice in Europe, and it will embody a new direction in the initiatives of

---

27. From the Latin "sacer" (sacred) and "facio" (I make).

the Ecumenical Patriarchate. Whereas the orphanage was at one time forcibly closed by Turkish authorities in an act of religious intolerance, it is highly expected to be returned to the Ecumenical Patriarchate through a just process in the European Court of Human Rights,[28] which ruled in favor of returning this historic property of the Ecumenical Patriarchate. The purpose of the center will be to translate theory into practice, providing educational resources to advance ecological transformation and interfaith tolerance.

The center will focus on climate change and the related changes needed in human behavior and ethics. It will serve as a source of inspiration and awareness for resolving religious issues related to the environment and peace, in cooperation with universities and policy centers on both local and international levels.

Dear friends, as we humbly learned very early on, and as we have repeatedly stressed throughout our ministry over the last twenty years, the environment is not only a political issue; it is also—indeed, it is primarily—a spiritual issue. Moreover, it directly affects all of us in the most personal and the most tangible manner. We can no longer afford to be passive observers in this crucial debate. In 2002, at the conclusion of the Adriatic Symposium, together with His Holiness, the late Pope John Paul II, we signed a declaration in Venice that proclaimed in optimism and prayer. Our conclusion was that:

It is not too late. God's world has incredible healing powers. Within a single generation, we could steer the earth toward our children's future. Let that generation start now.[29]

Indeed, let it start now. May God bless all of you.

ARTICLES

Guest Contribution to the International Journal of Heritage Studies

THINE OWN FROM THINE OWN . . .

As we contemplate the title of your esteemed journal and respond to the gracious invitation to contribute certain introductory remarks on our

---

28. After a unanimous decision of the European Court of Human Rights in 2008, the title of the orphanage was officially returned and presented to Ecumenical Patriarch Bartholomew on November 29, 2010.

29. See Chapter 9, "Declarations and Statements."

concern for the protection of the natural environment, it becomes immediately apparent that the word "heritage" in many ways crystallizes the very essence of the Orthodox theological and spiritual approach toward creation. [30] For, the term "heritage" indicates the fact that the world we enjoy comprises a gift we have received and not some property we own. Moreover, something inherited must be preserved and conveyed to generations that succeed us and not wasted selfishly without concern for those who come after us. Therefore, "heritage" is accompanied by a sense of respect and responsibility, which resembles the notion of tradition in religious circles.

Just as the word "heritage" bears a significant and symbolical dimension, so too the word "ecology" contains the prefix "eco," which derives from the Greek word *oikos*, signifying "home" or "dwelling." How unfortunate, then, and indeed how selfish it is that we have reduced the meaning and restricted the application of this crucial word. For, this world is indeed our home. Yet it is also the home of everyone, as it is the home of every animal creature, as well as of every form of life created by God. It is a sign of arrogance to presume that we human beings alone inhabit this world. Moreover, it is a sign of arrogance to imagine that only the present generation inhabits this earth.

The above-cited words from the Divine Liturgy attributed to our fourth-century predecessor and Archbishop of Constantinople, St. John Chrysostom, symbolize our conviction that this world is the fruit of divine generosity and boundless grace as well as of our commitment to respond with gratitude by respecting and protecting the natural environment—or, as we are commanded in Scripture, by "serving and preserving the earth" (Gen. 2.15) for the sake of future generations.

CREATION AS DIVINE GIFT

"Gift" (*doron*, in Greek) and "gift-in-return" (*antidoron*, in Greek) are liturgical terms that define our Orthodox theological understanding of the environmental question in a concise and clear manner. On the one hand, the natural environment comprises the unique *doron* of the Triune God to humankind. On the other hand, the appropriate *antidoron* of

---

30. Lead article in the journal, which appeared in 2006.

humankind toward its divine Maker is precisely the respect for and preservation of this gift, as well as its responsible and proper use. Each believer is called to celebrate life in a way that reflects the words of the Divine Liturgy: "Thine own [gifts] from Thine own we offer to Thee, in all and through all."

Thus the Eastern Orthodox Church proposes a liturgical worldview. It proclaims a world richly imbued by God and a God profoundly involved in this world. Our original sin, so it might be said, does not lie in any legalistic transgression that might incur divine wrath or human guilt. Instead, it lies in our stubborn refusal as human beings to receive the world as a gift of reconciliation with our planet and to regard the world as a sacrament of communion with the rest of humanity.

This is the reason why the Ecumenical Patriarchate has initiated and organized a number of international and inter-disciplinary symposia over the last decade: in the Aegean Sea (1995) and the Black Sea (1997), along the Danube River (1999) and in the Adriatic Sea (2002), in the Baltic Sea (2003) and, most recently, on the Amazon River (2006). For, like the air that we breathe, water is the very source of life; if defiled or despoiled, the element and essence of our existence is threatened. Put simply, environmental degradation and destruction is tantamount to suicide. One of the hymns of the Orthodox Church, chanted on the day of Christ's baptism[31] in the Jordan River, a feast of renewal and regeneration for the entire world, states: "I have become . . . the defilement of the air and the land and the water."

At a time when we have polluted the air that we breathe and the water that we drink, we are called to restore within ourselves a sense of awe and delight, to respond to matter as to a mystery of ever-increasing connections.

As a gift from God to humanity, creation becomes our companion, given to us for the sake of living in harmony with it and with others. We are to use its resources in measure, to cultivate it in love, and to preserve it in accordance with the Scriptural command (cf. Gen. 2.15). Within the unimpaired natural environment, humanity discovers deep spiritual peace and rest; and in humanity that is spiritually cultivated by the peaceful grace of God, nature recognizes its harmonious and rightful place.

---

31. Celebrated on January 6.

Nevertheless, the first-created human being misused freedom, preferring alienation from God-the-Giver and attachment to God's gift. Consequently, the double relationship of humanity to God and creation was distorted and humanity became preoccupied with using and consuming the earth's resources. In this way, the human blessedness derived from the love between God and humanity ceased, and humanity sought to fill this void by drawing from creation the blessedness that was lacking. From grateful user, the human person became greedy abuser. In order to remedy this situation, human beings are called to be "eucharistic" and "ascetic," namely to be thankful by offering glory to God for the gift of creation, while at the same time being respectful by practicing responsibility in the web of creation.

EUCHARISTIC AND ASCETIC BEINGS

Let us reflect further on these two words "eucharistic" and "ascetic." The implications of the first word are quite easily appreciated. In calling for a "eucharistic spirit," the Orthodox Church is reminding us that the created world is not simply our possession, but rather it is a gift—a gift from God the Creator, a healing gift, a gift of wonder and beauty. Therefore, the proper response, upon receiving such a gift, is to accept and embrace it with gratitude and thanksgiving. This is surely a distinctive characteristic of human beings. Humankind is not merely a logical or a political animal. Above all, human beings are eucharistic animals, capable of gratitude and endowed with the power to bless God for the gift of creation. Other animals express their gratefulness simply by being themselves, by living in the world through their own instinctive manner; yet we human beings possess self-awareness in an intuitive manner, and so consciously and by deliberate choice we can thank God with eucharistic joy. Without such thanksgiving, we are not truly human.

A eucharistic spirit implies using the earth's natural resources with thankfulness, offering them back to God—indeed, not only them, but also ourselves. In the Sacrament of the Eucharist, we return to God what is His own: namely, the bread and the wine, together with the entire community. All of us and all things represent the fruits of creation, which are no longer imprisoned by a fallen world, but returned as liberated, purified from their fallen state, and capable of receiving the divine presence within themselves. Whoever gives thanks also experiences the joy that comes from the appreciation of that for which he or she is thankful. Conversely, whoever does not feel the need to be thankful for the wonder and beauty of the world, but instead demonstrates only selfishness or indifference, can never experience a deeper, divine joy, but only sullen and inhumane satisfaction.

Second, we have the "ascetic ethos" of Orthodoxy that involves fasting and other similar spiritual disciplines. These make us recognize that all things we take for granted are, in fact, God's gifts provided in order to satisfy our needs as they are shared fairly among all people. However, they are not ours to abuse and waste simply because we have the desire to consume them or the ability to pay for them.

The "ascetic ethos" is the intention and discipline to protect the gift and to preserve nature harmless. It is the struggle for self-control, whereby we no longer willfully consume every fruit, but instead manifest a sense of self-restraint and abstinence from certain fruits. Both the protection and the self-restraint are expressions of love for all of humanity and for the entire natural creation. Such love alone can protect the world from unnecessary waste and inevitable destruction. After all, just as the true nature of "God is love" (1 Jn 4.8), so, too, humanity is originally and innately endowed with love.

Our purpose is thus enjoined to the priest's prayer in the Divine Liturgy: "In offering to Thee, Thine own [gifts] from Thine own, in all and through all—we praise Thee, we bless Thee, and we give thanks to Thee, O Lord." Then, we are able to embrace all people and all things—not with fear or necessity, but with love and joy. Then, we learn to care for the plants and for the animals, for the trees and for the rivers, for the mountains and for the seas, for all human beings, and for the whole natural environment. Then, we discover joy—rather than inflicting sorrow—in our life and in our world. Then, we create and promote instruments of peace and life, not tools of violence and death. Then, creation on the one hand and humanity on the other hand—the one that encompasses and the one that is encompassed—correspond fully and cooperate with one another. For, they are no longer in contradiction or in conflict or in competition. Then, just as humanity offers creation in an act of priestly service and sacrifice to God, so also does creation offer itself in return as a gift to humanity and to the generations that are to follow. Then, everything becomes a form of exchange, the fruit of abundance, and a fulfillment of love. Then, everything celebrates—what St. Maximus the Confessor in the seventh century called—a "cosmic liturgy."

A NEW WORLDVIEW

The crisis that we are facing in our world is not primarily ecological. It is a crisis concerning the way we envisage or imagine the world. We are treating our planet in an inhuman, godless manner precisely because we fail to see it as a gift inherited from above; it is our obligation to receive, respect, and render this gift to future generations. Therefore, before we can effectively deal with problems of our environment, we must change the way we envisage the world. Otherwise, we are simply dealing with symptoms, not with their causes. We require a new worldview if we are to desire "a new earth" (Rev. 21.1).

Therefore, let us acquire a "eucharistic spirit" and an "ascetic ethos" bearing in mind that everything in the natural world, whether great or small, has its importance within the universe and for the life of the world; nothing whatsoever is useless or contemptible. Let us regard ourselves as responsible before God for every living creature and for all the natural creation; let us treat everything with proper love and utmost care. Only in this way shall we secure a physical environment where life for the coming generations of humankind will be healthy and happy. The unquenchable greed of our generation constitutes a mortal sin inasmuch as it results in destruction and death. This greed in turn leads to the deprivation of our children's generation, in spite of our desire to bequeath to them a better future. Ultimately, it is for our children that we must perceive our every action in the world as having a direct effect upon the future of the environment.

As we declared some years ago in Venice with the late Pontiff of the Roman Catholic Church, Pope John Paul II:

> It is not too late. God's world has incredible healing powers. Within a single generation, we could steer the earth toward our children's future. Let that generation start now, with God's help and blessing.[32]

We must frankly admit that humankind is entitled to something better than what we see around us. We and, much more, our children and future

---

32. See Chapter 9, "Declarations and Statements."

generations, are entitled to a better world, a world free from degradation, violence, and bloodshed, a world of generosity and love. It is selfless and sacrificial love for our children that will show us the path that we must follow into the future.

Article for Seminarium, *Vatican, May 2010*

READING THE BOOK OF NATURE

In the late third century, St. Anthony of Egypt,[33] the "father of monasticism," described nature as a book that teaches us about the beauty of God's creation: "My book is the nature of creation; therein, I read the works of God."[34] The extraordinary spiritual collection entitled *The Philokalia*[35] records St. Anthony as saying: "Creation declares in a loud voice its Maker and master." This is how Orthodox theology and spirituality perceive the natural environment. There is, as St. Maximus the Confessor would claim in the seventh century, a liturgical or sacramental dimension to creation. The whole world, as he observed, is a "cosmic liturgy." For, as St. Maximus observes: "Creation is a sacred book, whose letters and syllables are the universal aspects of creation; just as Scripture is a beautiful world, which is constituted of heaven and earth and all that lies in between." What, then, is the Orthodox theological and liturgical vision of the world?

As a young child, as we accompanied the priest of our local village to services in remote chapels on my native island of Imvros in Turkey, the connection of the beautiful mountainside to the splendor of liturgy was evident. This is because the natural environment provides a broader, panoramic vision of the world. In general, nature's beauty leads to a more

---

33. Anthony the Great (251–356) is considered the pioneer and founder of monasticism in Egypt.

34. Entitled "And God Saw That Everything Was Good." The journal is published by the Congregation for Catholic Education. An abridged version of this text, entitled "The World as Sacrament," was delivered as an address to doctoral students of the Saints Cyril and Methodius Postgraduate School in Moscow (May 26, 2010).

35. An anthology of texts on prayer written between the fourth and fifteenth centuries, originally compiled in the eighteenth century by St. Nikodemus of the Holy Mountain (1749–1809) and St. Makarios of Corinth (1731–1805).

open view of the life and created world, somewhat resembling a wide-angle focus from a camera, which ultimately prevents us human beings from using or abusing its natural resources in a selfish, narrow-minded way.

In order, however, to reach this point of maturity and dignity toward the natural environment, we must take the time to listen to the voice of creation. And in order to do this, we must first be silent. The virtue of silence is perhaps the most valuable human quality underlined in *The Philokalia*. Indeed, silence is a fundamental element, which is critical in developing a balanced environmental ethos as an alternative to the ways that we currently relate to the earth and deplete its natural resources. *The Sayings of the Desert Fathers*[36] relate of Abba Chaeremon that, in the fourth century, he deliberately constructed his cell "forty miles from the church and ten miles from the water" so that he might struggle a little to do his daily chores. In Turkey today, the Princes Island of Heybeliada (or Halki) still forbids the traffic of cars.

So if we are silent, we will learn to appreciate how "the heavens declare the glory of God, and the firmament proclaims the creation of His hands" (Ps. 19.1). The ancient Liturgy of St. James, celebrated only twice a year in Orthodox Churches, affirms the same conviction:

> The heavens declare the glory of heaven; the earth proclaims the sovereignty of God; the sea heralds the authority of the Lord; and every material and spiritual creature preaches the magnificence of God at all times.

When God spoke to Moses in the burning bush, communication occurred through a silent voice, as St. Gregory of Nyssa informs us in his mystical classic, *The Life of Moses*. St. Gregory believes that we can discern God's presence simply by gazing at and listening to creation. Therefore, nature is a book opened wide for all to read and to learn. Each plant, each animal, and each micro-organism tells a story, unfolds a mystery, relates an extraordinary harmony and balance, which are interdependent and complementary. Everything points to the same encounter and mystery.

---

36. A fifth-century anthology of anecdotes from the early monastics. See, for example, Benedicta Ward, ed., *The Sayings of the Desert Fathers: The Alphabetical Collection* (London: Mowbrays, 1975).

The same dialogue of communication and mystery of communion is detected in the galaxies, where the countless stars betray the same mystical beauty and mathematical inter-connectedness. We do not need this perspective in order to believe in God or to prove His existence. We need it to breathe; we need it for us simply to be. The coexistence and correlation between the boundlessly infinite and the most insignificantly finite things in our world articulate a concelebration of joy and love.

It is unfortunate that we lead our life without even noticing the environmental concert that is playing out before our very eyes and ears. In this orchestra, each minute detail plays a critical role, and every trivial aspect participates in an essential way. No single member—human or other—can be removed without the entire symphony being affected. No single tree or animal can be removed without the entire picture being profoundly distorted, if not destroyed. When will we begin to learn and teach the alphabet of this divine language, so mysteriously concealed in nature?

ORTHODOX THEOLOGY AND THE NATURAL ENVIRONMENT

In its foremost symbol and declaration of faith, known as the Nicene-Constantinopolitan Creed, the Orthodox Church confesses "one God, maker of heaven and earth, and of all things visible and invisible." Thus, an Orthodox Christian perspective on the natural environment derives from the fundamental belief that the world was created by God. The Judaeo-Christian Scriptures state that "God saw everything that was created and, indeed, it was very good" (Gen. 1.31). Indeed, the Greek word for "good" signifies beauty and not merely goodness; the world was created "beautiful" by a loving Creator.

From this fundamental belief in the sacredness and beauty of all creation, the Orthodox Church articulates its crucial concept of cosmic transfiguration. This emphasis of Orthodox theology on personal and cosmic transfiguration is especially apparent in its liturgical feasts. The Feast of Christ's Transfiguration[37] highlights the sacredness of all creation, which receives and offers a foretaste of the final resurrection and restoration of all things in the age to come. The fifth-century *Macarian*

---

37. Celebrated on August 6.

Homilies[38] underline this connection between the Transfiguration of Christ and the sanctification of human nature:

> Just as the Lord's body was glorified, when he went up [Mt. Tabor] and was transfigured into glory and into infinite light . . . so, too, our human nature is transformed into the power of God, being kindled into fire and light.[39]

Yet the hymns of the day extend this divine light and transformative power to the whole world:

> Today, on Mt. Tabor, in the manifestation of your light, O Lord, You were unaltered from the light of the unbegotten Father. We have seen the Father as light, and the Spirit as light, guiding with light the entire creation.

Moreover, the Feast of the Baptism of Jesus Christ on January 6 is known as the Theophany (meaning "the epiphany" or "revelation of God") because it manifests the perfect obedience of Christ to the original command of Genesis and restores the purpose of the world as it was created and intended by God. The hymns of that day proclaim:

> The nature of waters is sanctified, the earth is blessed, and the heavens are enlightened . . . so that by the elements of creation, and by the angels, and by human beings, by things both visible and invisible, God's most holy name may be glorified.

The breadth and depth of the Orthodox cosmic vision implies that humanity is a part of this "theophany," which is always greater than any one individual. Of course, the human race plays a unique role and has a unique responsibility; but it nevertheless is only *a part of* the universe, which cannot be considered or conceived *apart from* the universe. As St. Maximus would say: "Human beings are not isolated from the rest of creation. They are bound, by their very nature, to the whole of creation." In this way, the natural environment ceases to be something that we

---

38. A series of fifty, or more, spiritual homilies written in Greek toward the end of the fourth and the beginning of the fifth centuries.

39. *Homily* XV.

observe objectively and exploit selfishly, and becomes a part of the "cosmic liturgy" or celebration of the essential interconnection and interdependence of all things. In light of this, in his twentieth-century classic *The Brothers Karamazov*, Fyodor Dostoevsky (1821–1881) urges:

> Love all God's creation, the whole of it and every grain of sand. Love every leaf, every ray of God's light. Love the animals, love the plants, love everything. If you love everything, you will perceive the divine mystery in things.[40]

According to divine revelation, the material and natural creation was granted by God to humanity as a gift, with the command to "serve and preserve the earth" (Gen. 2.15). Nevertheless, the first-created human being misused the gift of freedom, instead preferring alienation from God-the-Giver and attachment to God's gift. Consequently, the double relationship of humanity to God and creation was distorted and humanity became preoccupied with using and consuming the earth's resources. In this way, the human blessedness, which flows from the love between God and humanity, ceases to exist, and humanity sought to fill this void by drawing from creation itself—instead of from its Creator—the blessedness that was lacking. From grateful user, then, the human person became greedy abuser. In order to remedy this situation, human beings are called to return to a "eucharistic" and "ascetic" way of life, namely to be thankful by offering glory to God for the gift of creation, while at the same time being respectful by practicing responsibility within the web of creation.

EUCHARISTIC AND ASCETIC SPIRITUALITY

It is helpful to reflect further on these two critical terms: "eucharistic" and "ascetic." Both words are theological in nature, and the Church is called to teach these to its students and preach them to its faithful. The implications of the first word are quite easily appreciated. The term derives from the Greek word *eucharistia*, meaning "thanks," and is understood also as the deeper essence of liturgy. In the Orthodox Church, the Divine Liturgy is also called "the Sacred Eucharist." In calling for a "eucharistic spirit," then, the Orthodox Church is reminding us that the

---

40. See note 24.

created world is not simply our possession or property, but rather it is a treasure or gift—a gift from God the Creator, a healing gift, a gift of wonder and beauty. Therefore, the proper response, upon receiving such a gift, is to accept and embrace it with gratitude and thanksgiving.

Thanksgiving underlines the sacramental worldview of the Orthodox Church. From the very moment of creation, this world was offered by God as a gift to be transformed and returned in gratitude. This is precisely how the Orthodox spiritual way avoids the problem of the world's domination by humanity. For if this world is a sacred mystery, then this in itself precludes any attempt at mastery by human beings. Indeed, the mastery or exploitative control of the world's resources is identified more with Adam's "original sin" than with God's wonderful gift. It is the result of selfishness and greed, which arise from alienation from God and the abandonment of a sacramental worldview. Sin separated the sacred from the secular, dismissing the latter to the domain of evil and surrendering it as prey to exploitation.

Thanksgiving, then, is a distinctive and definitive characteristic of human beings. Humankind is not merely a logical or political being. Above all, human beings are eucharistic creatures, capable of gratitude and endowed with the power to bless God for the gift of creation. Again, the Greek word for "blessing" (*eulogia*) implies having a good word to say about something or someone; it is the opposite of cursing the world. Other animals express their gratefulness simply by being themselves, by living in the world through their own instinctive manner. Yet we human beings possess a sense of self-awareness in an intuitive manner, and so consciously and by deliberate choice we can thank God for the world with eucharistic joy. Without such thanksgiving, we are not truly human.

A eucharistic spirit also implies using the earth's natural resources with a spirit of thankfulness, offering them back to God with a sense of appreciation; indeed, we are to offer not only the earth's resources but ourselves. In the sacrament of the Eucharist, we return to God what is His own: namely, the bread and the wine, together with and through the entire community, which itself is offered in humble thanks to the Creator. As a result, God transforms the bread and wine, namely the world, into a mystery of encounter. All of us and all things represent the fruits of creation, which are no longer imprisoned by a fallen world, but returned as liberated, purified from their fallen state, and capable of receiving the divine presence within themselves.

Whoever gives thanks also experiences the joy that comes from appreciating that for which he or she is thankful. Conversely, whoever does not feel the need to be thankful for the wonder and beauty of the world, instead demonstrating selfishness or indifference, can never experience a deeper or divine joy, but only sullen sorrow and unquenched satisfaction. Such a person not only curses the world but experiences the world as curse. This is why people with so much can be so bitter, while others with so little can be so grateful.

The second term describing the proper human response to God's gift of creation is "ascetic," which derives from the Greek verb *askeo*, which implies a working of raw material with training or skill. Thus, we have the "ascetic ethos" of Orthodoxy that encourages fasting and other similar spiritual disciplines. These practices make us recognize that all that we take for granted are divine gifts, which are provided in order to satisfy our needs as they are shared fairly among all people. However, they are not ours to abuse and waste simply because we have the desire to consume them or the ability to pay for them.

The "ascetic ethos" is the disciplined effort to protect the gift of creation and to preserve nature intact. It is the struggle for self-restraint and self-control, whereby we no longer willfully consume every fruit, but instead manifest a sense of frugality and abstinence from certain fruits for the sake of valuing all fruits. Both the protection and the self-restraint are expressions of love for all of humanity and for the entire natural creation. Such love alone can protect the world from unnecessary waste and inevitable destruction. After all, just as the true nature of "God is love" (1 Jn 4.8), so, too, humanity is originally and innately endowed with the task of loving.

Our purpose is thus conjoined to the priest's prayer in the Divine Liturgy: "In offering to Thee, Thine own [gifts] from Thine own, in all and for all—we praise Thee, we bless Thee, and we give thanks to Thee, O Lord." Then, we are able to embrace all people and all things—not with fear or necessity, but with love and joy. Then, we learn to care for the plants and for the animals, for the trees and for the rivers, for the mountains and for the seas, for all human beings and for the whole natural environment. Then, we discover joy—rather than inflicting sorrow—in our life and in our world. As a result, we create and promote instruments of peace and life, not tools of violence and death. Then, creation on the one hand and humanity on the other hand—the one that encompasses and the one that is encompassed—correspond fully and cooperate with one another. For, they are no longer in contradiction or in conflict or in competition. Then, just as humanity offers creation in an act of priestly service and sacrifice, returning them to God, so also does creation offer itself in return as a gift to humanity for all generations that follow. Then, everything becomes a form of exchange, the fruit of abundance, and a fulfillment of love. Then, everything assumes its original vision and purpose, as God intended it from the moment of creation.

TEACHING THE DAYS OF CREATION

The brief, yet powerful statement found in Genesis, chapter 1 (verse 11), corresponds to the majesty of creation as understood in Orthodox theology, liturgy, and spirituality:

> Then God said: "Let the earth bring forth vegetation: plants yielding seed, and fruit trees of every kind on earth that bear fruit with the seed in it." And it was so. . . . And God saw that it was good. And there was evening and there was morning, the third day.

We all know the healing and nourishing essence of plants; we all appreciate their manifold creative and cosmetic usefulness.

> Consider the lilies, how they grow: they neither toil nor spin; yet I tell you, even Solomon in all his glory was not clothed like one of these." (Lk 12.27)

Even the humblest and lowliest manifestations of God's created world comprise the most fundamental elements of life and the most precious aspects of natural beauty.

Nevertheless, by overgrazing or deforestation, we tend to disturb the balance of the plant world. Whether by excessive irrigation or urban construction, we interrupt the magnificent epic of the natural world. Our selfish ways have led us to ignore plants, or else to undervalue their importance. Our understanding of plants is sparse and selective. Our outlook is greed-oriented and profit-centered.

However, plants are the center and source of life. Plants permit us to breathe and to dream. Plants provide the basis of spiritual and cultural life. A world without plants is a world without a sense of beauty. Indeed, a world without plants and vegetation is inconceivable and unimaginable. It would be the contradiction of life itself, tantamount to death. There is no such thing as a world where unsustainable development continues without critical reflection and self-control; there is no such thing as a planet that thoughtlessly and blindly proceeds along the present route of global warming. There is only wasteland and destruction. To adopt any other excuse or pretext is to deny the reality of land, water, and air pollution.

Plants are also the wisest of teachers and the best of models. For they turn toward light. They yearn for water. They cherish clean air. Their roots dig deep, while their reach is high. They are satisfied and sustained with so little. They transform and multiply everything that they draw from nature, including some things that appear wasteful or useless. They adapt spontaneously and produce abundantly—whether for the nourishment or admiration of others. They enjoy a microcosm of their own, while at the same time equally contributing to the macrocosm around them.

On the final days of creation, God is said to have made the variety of animals, as well as created man and woman in the divine image and likeness (Gen. 1.26). What most people seem to overlook is that the sixth day of creation is not entirely dedicated to the forming of Adam out of the earth. That sixth day was, in fact, shared with the creation of numerous "living creatures of every kind; cattle and creeping things and wild animals of the earth of every kind" (Gen. 1.24). This close connection between humanity and the rest of creation, from the very moment of genesis, is surely an important and powerful reminder of the intimate relationship that we share as human beings with the animal kingdom. While there is undoubtedly something unique about human creation in the divine image, there is more that unites us than separates us, not only as human beings but also with the created universe. It is a lesson we have learned in recent decades; but it is a lesson that we learned the hard way.

The saints of the early Eastern Church taught this same lesson long ago. They knew that a person with a pure heart was able to sense the connection with the rest of creation, especially the animal world. This is surely a reality that finds parallels in both Eastern and Western Christianity: One may recall Seraphim of Sarov (1759–1833) feeding the bear in the forest of the north, or Francis of Assisi (1181–1226) addressing the elements of the universe. The connection is not merely emotional; it is profoundly spiritual in its motive and content. It gives a sense of continuity and community with all of creation, while providing an expression of identity and compassion with it—a recognition that, as St. Paul put it, all things were created in Christ and in Christ all things hold together (Col. 1.15–17). This is why Abba Isaac of Nineveh can write from the desert of Syria:

> What is a merciful heart? It is a heart, which is burning with love for the whole of creation: for human beings, for birds, for beasts, for demons—for all of God's creatures. When such persons recall or regard these creatures, their eyes are filled with tears. An overwhelming compassion makes their heart grow small and weak, and they cannot endure to hear or see any kind of suffering, even the smallest pain, inflicted upon any creature. Therefore, these persons never cease to pray with tears even for the irrational animals, for the enemies of truth, as well as for those who do them evil, asking that these may be protected and receive God's mercy. They even pray for the reptiles with such great compassion, which rises endlessly in their heart until they shine again and are glorious like God.[41]

Thus, love for God, love for human beings, and love for animals cannot be separated sharply. There may a hierarchy of priority, but it is not a sharp distinction of comparison. The truth is that we are all one family—human beings and the living world alike—and all of us look to God the Creator: "These all look to you to give them . . . When you open your hand, they are filled with good things. When you hide your face, they are dismayed. When you take away their breath, they die and return to their dust" (Ps. 104.28–29).

INTERPRETING THE CONCEPT OF SIN

If the earth is sacred, our relationship with the natural environment is mystical or sacramental; that is to say, it contains the seed and trace of God. In many ways, then, the "sin of Adam" is precisely his refusal to

---

41. *Ascetic Treatises*, Homily 48.

receive the world as a gift of encounter and communion with God and with the rest of creation. St. Paul's Letter to the Romans emphasizes the consequences of the Fall, that "from the beginning till now, the entire creation, which as we know has been groaning in pain" (Rom. 8.22), also "awaits with eager longing this revelation by the children of God" (Rom. 8.19).

However, far too long have we focused—as churches and as religious communities—on the notion of sin as a rupture in individual relations either with each other or between humanity and God. The environmental crisis that we are facing reminds us of the cosmic proportions and consequences of sin, which are more than merely social or narrowly spiritual. It is our conviction that every act of pollution or destruction of the natural environment is an offense against God as Creator.

We are, as human beings, responsible for creation; but we have behaved as if we own creation. The problem of the environment is primarily neither an ethical nor a moral issue. It is an ontological issue, demanding a new way of being as well as a new way of behaving. Repentance implies precisely a radical change of ways, a new outlook and vision. The Greek word for repentance is *metanoia*, which signifies an inner transformation that inevitably involves a change in one's entire worldview. We repent not simply for things we feel that we do wrongly against God. Furthermore, we repent not simply for things that make us guilty in our relations with other people. Rather, we repent for the way we regard the world and, therefore, invariably treat—in fact, mistreat—the world around us.

This is why healing a broken environment is a matter of truthfulness to God, humanity, and the created order. We are called to broaden the traditional concept of sin—beyond individual and social implications—to include environmental damage! Some fifteen years ago, at a conference in Santa Barbara, we declared:

> To commit a crime against the natural world is a sin. For human beings to cause species to become extinct and destroy the biological diversity of God's creation . . . to degrade the integrity of the earth by causing climate change . . . to strip the earth of its natural forests, or destroy its wetlands . . . to contaminate the earth's waters, its land, its air, and its life—all of these are sins. [42]

42. November 8, 1997. For the full text, see the third section of this chapter, "Conferences and Events."

In this respect, the concept of sin must be broadened to include all human beings and all of created nature. Religions must become sensitized to the seriousness and implications of this kind of sin if they are to encourage the right values and inspire the necessary virtues to protect God's creation in its human, animal, and natural expressions. During the international negotiations, which took place at the Hague in 2000,[43] we strongly emphasized the threat to our planet's fragile ecosystems posed by global warming, as well as the urgent need for all religions to underline the need for renewed repentance in our attitude toward nature.

SOCIAL AND POLITICAL IMPLICATIONS

Orthodox theology takes all of this a step further step and recognizes the natural creation as inseparable from the identity and destiny of humanity because every human action leaves a lasting imprint on the body of the earth. Human attitudes and behavior toward creation directly impact and reflect human attitudes and behavior toward other people. Ecology is inevitably related in both its etymology and meaning to economy; our global economy is simply outgrowing the capacity of our planet to support it. At stake is not just our ability to live in a sustainable way, but our very survival. Scientists estimate that those most hurt by global warming in years to come will be those who can least afford it. Therefore, the ecological problem of pollution is invariably connected to the social problem of poverty; and so all ecological activity is ultimately measured and properly judged by its impact and effect upon other people, and especially the poor (see Matt. 25).

How, then, does respect for the natural environment translate into contemporary attitudes and action? The issue of environmental pollution and degradation cannot be isolated for the purpose of understanding or resolution. The environment is the home that surrounds the human species and comprises the human habitat. Therefore, the environment cannot be appreciated or assessed alone, without a direct connection to the unique creature that it surrounds, namely humanity. Concern for the environment implies also concern for human problems of poverty, thirst,

---

43. The COP6 climate change convention organized by the United Nations in the Hague (November 13–24).

and hunger. This connection is detailed in a stark manner in the Parable of the Last Judgment, where the Lord says: "I was hungry and you gave me food; I was thirsty and you gave me something to drink" (Matt. 25.35).

Concern, then, for ecological issues is directly related to concern for issues of social justice, and particularly of world hunger. A Church that neglects to pray for the natural environment is a Church that refuses to offer food and drink to a suffering humanity. At the same time, a society that ignores the mandate to care for all human beings is a society that mistreats the very creation of God, including the natural environment. It is tantamount to blasphemy.

The terms "ecology" and "economy" share the same etymological root. Their common prefix "eco" derives from the Greek word *oikos*, which signifies "home" or "dwelling." It is unfortunate and selfish, however, that we have restricted the application of this word to ourselves, as if we are the only inhabitants of this world. The fact is that no economic system—no matter how technologically or socially advanced—can survive the collapse of the environmental systems that support it. This planet is indeed our home; yet it is also the home of everyone, as it is the home of every animal creature, as well as of every form of life created by God. It is a sign of arrogance to presume that we human beings alone inhabit this world. Indeed, by the same token, it is also a sign of arrogance to imagine that only the present generation inhabits this earth.

Therefore, as one of the more serious ethical, social, and political problems, poverty is directly and deeply connected to the ecological crisis. A poor farmer in Asia, in Africa, or in North America will daily face the reality of poverty. For these persons, the misuse of technology or the eradication of trees is not merely harmful to the environment or destructive of nature; rather, it practically and profoundly affects the very survival of their families. Terminology such as "ecology," "deforestation," or "over-fishing" is entirely absent from their daily conversation or concern. The "developed" world cannot demand from the "developing" poor an intellectual understanding with regard to the protection of the few earthly paradises that remain, especially in light of the fact that less than 10 percent of the world's population consumes over 90 percent of the earth's natural resources. However, with proper education, the "developing" world would be far more willing than the "developed" world to cooperate for the protection of creation.

Closely related to the problem of poverty is the problem of unemployment, which plagues societies throughout the world. It is abundantly clear that neither the moral counsel of religious leaders nor fragmented measures by socio-economic strategists or political policymakers could be sufficient to curb this growing tragedy. The problem of unemployment compels us to re-examine the priorities of affluent societies in the West, and especially the unrestricted advance of development. We appear to be trapped in the tyrannical cycle created by a need for constant productivity rises and increases in the supply of consumer goods. However, placing these two "necessities" on an equal footing imposes on society a relentless need for unending perfection and growth, while restricting power over production to fewer and fewer. Concurrently, real or imaginary consumer needs constantly increase and rapidly expand. Thus the economy assumes a life of its own, a vicious cycle that becomes independent of human need or human concern. What is needed is a radical change in politics and economics, one that underscores the unique and primary value of the human person, thereby placing a human face on the issues of employment and productivity.

AN OPTIMISTIC WORLDVIEW

We have repeatedly stated that the crisis that we are facing in our world is not primarily ecological. It is a crisis concerning the way we envisage or imagine the world. We are treating our planet in an inhuman, godless manner precisely because we fail to see it as a gift inherited from above; it is our obligation to receive, respect, and in turn hand on this gift to future generations. Therefore, before we can effectively deal with problems of our environment, we must change the way we perceive the world. Otherwise, we are simply dealing with symptoms, not with their causes. We require a new worldview if we are to desire "a new earth" (Rev. 21.1).

So let us acquire a "eucharistic spirit" and an "ascetic ethos" bearing in mind that everything in the natural world, whether great or small, has its importance *within the universe and for the life of the world*; nothing whatsoever is useless or contemptible. Let us regard ourselves as responsible before God for every living creature and for the whole of natural creation; let us treat everything with proper love and utmost care. Only in this way shall we secure a physical environment where life for the coming generations of humankind will be healthy and happy. Otherwise, the unquenchable greed of our generation will constitute a mortal sin resulting in destruction and death. This greed in turn will lead to the deprivation of our children's generation, in spite of our desire and claim to bequeath to them a better future. Ultimately, it is for our children that we must perceive our every action in the world as having a direct effect upon the future of the environment.

This is the source of our optimism. As we declared some years ago in Venice (June 10, 2002) with the late Pontiff of the Roman Catholic Church, Pope John Paul II (1978–2005):

> It is not too late. God's world has incredible healing powers. Within a single generation, we could steer the earth toward our children's future. Let that generation start now, with God's help and blessing.[44]

The same sentiments were jointly communicated with the current Pope, Benedict XVI, during his official visit to the Ecumenical Patriarchate on November 30, 2006:

> In the face of the great threats to the natural environment, we wish to express our concern at the negative consequences for humanity and for the whole of creation, which can result from economic and technological progress that does not know its limits. As religious leaders, we consider it one of our duties to encourage and to support all efforts made to protect God's creation, and to bequeath to future generations a world in which they will be able to live.[45]

The natural environment—the forest, the water, the land—belongs not only to the present generation but also to future generations. We must frankly admit that humankind is entitled to something better than what we see around us. We and, much more, our children and future generations are entitled to a better and brighter world, a world free from degradation, violence and bloodshed, a world of generosity and love. It is selfless and sacrificial love for our children that will show us the path that we must follow into the future.

---

44. See Chapter 9, "Declarations and Statements."

45. See the second volume in this series, entitled *Speaking the Truth in Love* (New York: Fordham University Press, 2011).

CNN Guest Editorial, New York, October 19, 2010

SURVIVAL AND SALVATION

Last October, the Ecumenical Patriarchate convened an international, interdisciplinary, and interfaith symposium in New Orleans on the Mississippi River, the eighth in a series of high-level conferences exploring the impact of our lifestyle and consumption on our planet's major bodies of water. [46] Similar symposia have met in the Aegean and Black Seas, in the Adriatic and Baltic Seas, along the Danube and Amazon Rivers, and on the Arctic.

At first glance, it may appear strange for a religious institution concerned with "sacred" values to be so profoundly involved in "worldly" issues. After all, what does preserving the planet have to do with saving the soul? It is commonly assumed that global climate change and the exploitation of our nature's resources are matters that concern politicians, scientists, and technocrats. At best, perhaps, they are the preoccupation of special interest groups or naturalists.

So the preoccupation of the Orthodox Christian Church and, in particular, her highest spiritual authority, the Ecumenical Patriarchate, with the environmental crisis will probably come as a surprise to many people. Yet, there are no two ways of looking at either the world or God. There can be no double vision or worldview: one religious and the other profane, one spiritual and the other secular. In our worldview and understanding, there can be no distinction between concern for human welfare and concern for ecological preservation.

Nature is a book, opened wide for all to read and to learn, to savor and celebrate. It tells a unique story; it unfolds a profound mystery; it relates an extraordinary harmony and balance, which are interdependent and complementary. The way we relate to nature as creation directly reflects the way we relate to God as Creator. The sensitivity with which we handle the natural environment clearly mirrors the sacredness that we reserve for the divine. We must treat nature with the same awe and wonder that we

---

46. This opinion article was prepared for the online edition of CNN in light of the Hollister Award, which was presented to His All Holiness (*in absentia*) by the Temple of Understanding in New York City on October 19, 2010. Reprinted with permission. See http://edition.cnn.com/2010/OPINION/10/19/bartholomew.souls.plant/index.html.

reserve for human beings. And we do not need this insight in order to believe in God or to prove His existence. We need it to breathe; we need it for us simply to be.

At stake is not just our ability to live in a sustainable way, but our very survival. Scientists estimate that those most hurt by global warming in years to come will be those who can least afford it. Therefore, the ecological problem of pollution is invariably connected to the social problem of poverty; and so all ecological activity is ultimately measured and properly judged by its impact upon people, and especially its effect upon the poor.

In our efforts, then, to contain global warming, we are admitting just how prepared we are to sacrifice some of our greedy lifestyles. When will we learn to say: "Enough!"? When will we direct our focus away from what we want to what the world needs? When will we understand how important it is to leave as light a footprint as possible on this planet for the sake of future generations? We must choose to care. Otherwise, we do not really care at all.

We are all in this together. Indeed, the natural environment unites us in ways that transcend doctrinal differences. We may differ in our conception of the planet's origin, whether biblical or scientific. But we all agree on the necessity to protect its natural resources, which are neither limitless nor negotiable.

It is not too late to respond—as a people and as a planet. We could steer the earth toward our children's future. Yet we can no longer afford to wait; we can no longer afford not to act. People of faith must assume leadership in this effort; citizens of the world must clearly express their opinion; and political leaders must act accordingly. Deadlines can no longer be postponed; indecision and inaction are not options.

We are optimistic about turning the tide, quite simply because we are optimistic about humanity's potential. Let us not simply respond in principle; let us respond in practice. Let us listen to one another; let us work together; let us offer the earth an opportunity to heal so that it will continue to nurture us.

3

Nature and Cosmos

Beauty and Harmony of Creation

ENVIRONMENT AND CITY

We are deeply grateful for the honor bestowed on us today, an honor that we accept not on behalf of one individual, but on behalf of the Ecumenical Patriarchate and the entire Holy Orthodox Church in whose rich vineyard we are privileged to labor. We may toil in this vineyard, we may plant seeds here, we may harvest its fruits, but it is God's vineyard, they are God's seeds, and it is God's fruit. All glory, then, is due to Him.

The company within which we are receiving it increases the joy we feel in accepting this degree. For we are in the presence of His Royal Highness, Prince Philip, of our brother in Christ, the Archbishop of Canterbury, of the Lord Bishop of London, of the Lord Mayor of London, and, of course, of the administration, faculty, and students of this extraordinary university, which serves and enriches London, the largest city in Europe.

We warmly thank His Lordship the Bishop of Stepney for his most kind and generous introduction. Our joy is even further multiplied because this honorary degree is being conferred upon us on the momentous occasion of the centenary of City University. One hundred years is indeed a landmark achievement worthy of praise and recognition. We are deeply touched that you have chosen to include us in these celebrations. To you, the esteemed administration, the faculty, and beloved students, we extend our heartfelt congratulations and paternal prayers that God, the Giver of Light, may continue to illumine your hearts and minds as you increase in knowledge and wisdom. May God bless you with yet another centenary celebration.

The ancient Greeks believed that human beings could rise to their full potential only within the context of a city. This great university and its talented students are evidence of that ancient wisdom. Nevertheless, increasingly these days, we witness another, darker aspect of life in our cities. We observe children without clothing, food, or shelter; we see people without jobs; we hear of brothers killing brothers; we recognize broken families, broken lives, and broken dreams. Therefore, we ask ourselves: why? What went wrong? How can this be?

Our first instinct is to doubt that wisdom of the ancients. However, our better instinct is in fact to believe it even more. For, if we truly believe that cities offer great opportunities, we will be driven to discover why it is that so many people are not finding those opportunities. What is missing? What is lacking? The answer, we believe, lies in faith, not in knowledge or wealth or political action. It lies simply in faith. Knowledge expands the mind, but faith can open the heart. Wealth builds houses, but faith can move mountains. Politics does the possible, but faith can do the impossible.

Western civilization has brought about the greatest of human achievements—from medical miracles to people on the moon, from stable democracies to the high standards of living. Yet, these have come with a price, and that price is most evident on the streets of our cities. Politicians and professors alone cannot heal the problems of Western society, be they pornography, pollution, drugs, poverty, crime, war, or homelessness. Religious leaders have a central and inspirational role to play in raising the spiritual principles of love, tolerance, morality and renewal to the fore.

This is why we consider this degree such a special honor. For, a secular university is bestowing it on a spiritual institution, thus demonstrating that one is not antithetical to the other. Indeed, more than this, it brings our two worlds—of academia and ecclesia—closer together, and for this we are truly grateful to God.

We are convinced that our mission today—namely, that of bringing the healing power of the Holy Spirit to all the children of God—is more vital than ever. The spirituality of the Church offers a different sort of fulfillment than that which is offered by the secularism of modern life. Here, too, there is no antithesis.

The failure of anthropocentric ideologies has left a void in many people's lives. The frantic pursuit of the future has sacrificed the inner peace of the past. We need to regain our religious outlook. We must urgently counter the effects of secular humanism with the teaching of the Church on humanity and the natural world and elevate the pursuit of the temporal toward a healthy respect for the eternal by bringing the one into harmony with the other. Moreover, we must repair the torn fabric of society by reminding ourselves every day that the misfortune of some of us, affects the fortune of all of us.

Our society resembles the lawyer who asked Jesus: "Teacher, what must I do to inherit eternal life?" Jesus responded: "What is written in the law?" The lawyer said: "You shall love the Lord your God with all your heart, and with all your soul, and with all your mind; and your neighbor as yourself" (Lk 10.25–27). This led to a further question—one that is of extremely relevance to our world today. We are referring to the question: "Who is my neighbor?" (Lk 10.29).

Jesus answered with the story of a man who was robbed and beaten on the road leading from Jerusalem to Jericho. A priest approached that man, and—and, in much the same way as so many of us step over a homeless person today—crossed to the other side of the road. Then a Levite came by; he, too, avoided the situation by crossing the road. Yet a Samaritan who happened to be traveling down that road was moved to bind the man's wounds, take him to the closest inn, and tend to him. Jesus asked: "Which of these three, do you think, proved neighbor to the man who fell among robbers?" (Lk 10.36). When the lawyer chose the Good Samaritan, Jesus simply said: "Go and do likewise" (Lk 10.37).

Today, there is hardly a more important question than: "Who is my neighbor?" The future of our world rests on how we respond to this. Sadly, we do not always answer as we should. In Bosnia, where warfare still rages, like the priest in Jesus' parable, too many have crossed the road rather than confront the situation. In Los Angeles, in London and in St. Petersburg, too many of our children have been abandoned to the urban warfare of the streets. In South Africa, on the other hand, we have seen millions of our fellow human beings behave like the Good Samaritan. The South Africans are certainly proving themselves true neighbors.

If God would only grant us the power to plant just one idea, as though it were a seed, in the fertile minds that are gathered in this great cathedral today—in order, thus, to return the favor for this degree by offering back to this secular institution a simple, yet profound spiritual exhortation—it would be just this: "Go and do likewise." Know that every human being is your neighbor, and behave accordingly. Above all else, "Love your neighbor as yourself" (Lk 10.27).

SYMPOSIA

Opening Address, Symposium VI, Amazon River, July 14, 2006

THE FRAGILE BEAUTY OF THE WORLD

It is with great pleasure that we welcome you all, dignitaries and delegates, to the official opening of Symposium VI: Amazon, Source of Life. Since 1995, five water-borne symposia have been organized in major water bodies of Europe: the Aegean Sea, the Black Sea, the Danube River, the Adriatic Sea, and the Baltic Sea. The participation of regional and international representatives of the scientific, religious, and media worlds ensures that the message and outcomes of our symposium, with regard to crucial and specific issues, will be brought to the attention of the world community. The present symposium is honored by the joint patronage of His Excellency Kofi Annan, Secretary General of the United Nations.[1]

This symposium is in many ways both historical and unique. It is the first time that our initiatives have ventured beyond European boundaries but our gathering underlines—on a broader, global level—the critical role of the Amazon River for the future of our planet. This river comprises a microcosm of our planet. In its waters, we observe many of the world's ecological issues. We are humbled in its presence. We have come to listen to its story, to learn from its history, to admire its fragile beauty, and to gain hope for the entire world from its resilience.

---

1. A Ghanaian diplomat, who served as the seventh secretary general of the United Nations (1997–2006).

We are conscious of the consequences of human activity on the Brazilian rainforest. Environmental issues are very much at the forefront of daily news. We hear of air and water pollution, of global warming and the threatened extinction of numerous animal or plant species. The statistics are indeed alarming. How should we react? What do they mean for development and for the ways we are accustomed to living?

DIVINE ECONOMY AND THE ECONOMY OF NATURE

Every product we make and enjoy (from the paper we work with, to processed meat and the soy beans that sustain its industry), every tree we fell, every building we construct, every road we travel, definitively and permanently alter creation. At the basis of this alteration—or perhaps we should characterize it as abuse—of creation is a fundamental difference between human, natural, and divine economies. In the Orthodox tradition, the phrase "divine economy" is used to describe God's extraordinary acts of love and providence toward humanity and creation. "Economy" is derived from the Greek word *oikonomia*, which implies the management of an environment or household (*oikos*), which is also the root of the word "ecology" (*oikologia*). Let us consider, however, the radical distinction between the various kinds of economy. Our economy tends to use and discard; natural economy is normally cyclical and replenishes; God's economy is always compassionate and nurturing. Nature's economy is profoundly violated by our wasteful economy, which in turn constitutes a direct offence to the divine economy. The prophet Ezekiel again recognized this abuse of the natural ecosystems when he observed:

> Is it not enough to feed on good pasture? Must you also trample the rest with your feet? Is it not sufficient to drink clear water? Must you also muddy the rest with your feet? (34.18)

Our perspective is neither that of a scientist nor that of an economist; our principles are derived from the altar of the Church and the heart of theology. In this respect, the liturgy provides for us a mystical basis for a broader, spiritual worldview. It both reflects the way we respond to creation and molds the way we respect creation.

Perhaps no other place in the world reflects so apparently or records so articulately both the sacred beauty of creation and the consequences of human choices. A spiritual worldview should inform our concept of creation and define our conduct within this world. This worldview is neither a political plan nor an economic strategy. It is essentially a way of reflecting on what it means to perceive the world through the lens of the soul.

Let us consider our own presence on this great river. The question we must address to ourselves in all honesty is: Have we come here as pilgrims or as travelers? What have we come here to see?

THE VISION OF LITURGY

Seeing clearly is precisely what the liturgy teaches us to do. Our eyes are opened to see the beauty of created things. The world of the liturgy reveals the eternal dimension in all that we see and experience. It enables us to hear new sounds and behold new images as we travel along the Amazon River. It creates in us a mystical appreciation and genuine affection for everything that surrounds us. The truth is that we have been inexorably locked within the self-centered confines of our own individual concerns with no access to the world beyond us. We have violated the sacred covenant between our selves, our world, and our God.

The liturgy restores this covenant; it reminds us of another way and of another world. It offers a corrective to a wasteful, consumer culture that gives value only to the here and now. The liturgy converts the attentive person from a restricted, limited point of view to a fuller, spiritual vision "in Him through whom all things live, move, and have their being" (Acts 17.28). It provides for us another means of comprehension and communication. The liturgy is the eternal celebration of the fragile beauty of this world.

In practical terms, this would naturally imply a way of life that would be respectful of the divine presence in creation. We should not be blindfolded by personal interests, but be sensitive to the sacredness of every peninsula and every island, every river and every stream, every basin and every landscape. If we are guilty of relentless waste, it is because we have lost the spirit of liturgy and worship. We are no longer respectful pilgrims on this earth; we have been reduced to careless consumers or passing travelers. How tragic it would be, for us all as delegates of this symposium, if we were simply to pass through the Amazon, like the indifferent priest in the Parable of the Good Samaritan. We must be responsible and responsive citizens of the world; we must be careful and caring pilgrims in this land. If we are not, in fact, moved to compassion, bandaging the wounds of the earth, assuming personal care, and contributing to the painful costs, then we might easily be confronted with the question: *Which of these do you resemble: the Good Samaritan or the indifferent priest?*

The liturgy guides us to a life that sees more clearly and shares more fairly, moving away from what we want as individuals to what the world needs globally. This in turn requires that we move away from greed and control and gradually value everything for its place in creation and not simply its economic value to us, thereby restoring the original beauty of the world, seeing all things in God and God in all things.

MAKING CHOICES

Esteemed dignitaries and fellow participants, perhaps for the first time in the history of our world, we recognize that our decisions and choices immediately impact the environment. Today, we are able to direct our actions in a caring and compassionate way. It is up to us to shape our future; it is up to us to choose our destiny.

Breaking the vicious circle of ecological degradation is a choice with which we are uniquely endowed at this crucial moment in the history of our planet. This conference is a golden opportunity for us to recognize the unique role of every individual and every organization, in order that we may respect those more vulnerable in this situation, and in order that we may be prepared to assume responsibility for the health of our planet, an issue of critical significance and urgency.

As we officially declare the opening of Symposium VI: Amazon, Source of Life, may we all be inspired by grace and justice, guided by reason and responsibility, and filled with selflessness and compassion. May we be poised in the expectation to learn from the fragile beauty of God's creation, and from the unparalleled dynamism of the great Amazon River.

Homily on the 50th Anniversary, Santa Barbara Church, California, November 8, 1997

BEAUTY AND NATURE

This city combines the mountains, the forests, the valleys, and the waters, in a divine symphony praising their Creator. This church edifice stands with imposing symbolical dignity, at the top of this holy and sacred hill. Your altar is dedicated to the Great Martyr Barbara, whom we shall commemorate next month. She is renowned for her miraculous powers of healing the sense of sight. The vision of light is considered to be the most significant of human experiences. The sense of sight was regarded as the most profound sense in the classical Greek philosophical and patristic world. *Truth is beheld; it is not understood intellectually. God is seen; He is not examined theoretically. Beauty is perceived; it is not speculated abstractly.*

In the brief service just held, we were reminded of how, when God made the heavens and the earth, He said: "Let there be light." He made the waters, the land, the sky, and the living creatures. "And God saw that it was good [*kalon*]" (Gen. 1.4), which literally means "beautiful." This beauty is, first and foremost, the beauty of divine sacredness, a self-revelation and self-realization of God, who invites us to share in, and to enjoy that beauty. For, everything that lives and breathes is sacred and beautiful in the eyes of God. The whole world is a sacrament. The entire created cosmos is a burning bush of God's uncreated energies. And humankind stands as a priest before the altar of creation, as microcosm and mediator. Such is the true nature of things; or, as an Orthodox hymn describes it, "the truth of things," if only we have the eyes of faith to see it.

Realistically, we also know that this vision has been blurred; the image has been marred, by our sin. For we have presumed to control the order of things, and have therefore destroyed the hierarchy of creation. We have lost the dimension of beauty, and have come to a spiritual impasse where everything that we touch is invariably distorted or even destroyed. Nevertheless, through the divine Incarnation, our sight is once again restored, and we are once more enabled to discern the beauty of Christ's countenance "in all places of His dominion," and "in the least of our brothers and sisters" (Gen. 25.40). When "the Word became flesh" (Jn 1.14), we were endowed with new eyes, new ears, new senses altogether, in order to see the invisible in what is visible, and in order to experience the uncreated in what is created.

CREATION AND ICONS

This same truth is revealed in the iconography of our Church. They are a necessary expression of the world's sacredness. "Standing inside the

Church, we think that we are in heaven," as one hymn states, because the Church is the embodiment and sanctification of the whole world, while the created world is called to become the Church. The icons break down the wall of separation between the sacred and the profane, between this world and the next, between earth and heaven.

We are called today to rediscover this iconic dimension of creation. And it is this that distinguishes us as Christians and our awesome responsibility for the survival of our environment. In the Church, then, "through heaven and earth and sea, through wood and stone, through relics and Church buildings and the Cross, through angels and people, through all of creation, both visible and invisible, we offer veneration and honor to the Creator and master and Maker of all things, and to Him alone."[2]

All things are sacramental when seen in the light of God. In the Great Doxology that we just chanted, we prayed that "in His light we may see light, for He is the one who has shown us the light." And this beautiful building, with its traditional Byzantine architecture, and refreshing Mediterranean setting, is a symbol of that original beauty that was restored to us in Christ.

Let us open our eyes to see the world that God has made for us. Let us walk gently on the ground that so patiently tolerates our behavior. Then, as children of God, we can liberate the whole creation. Nothing less is expected of us. Nothing less dignifies us or the world around us. And nothing less is worthy of our high calling in Christ Jesus, who is adored and glorified with His eternal Father, and the all-holy, good and life-creating Spirit, now and ever, and unto the ages of ages.

EVENTS

Address at the Town Hall of Sydney, November 26, 1996

BEAUTY, NATURE, AND CITY

Sydney is the oldest, most populated, and most attractive city of the fifth continent; and it is a most significant center of social, financial, and cultural life in Australia. Yet, by the standards of the Old World, it is a

---

2. St. Leontius of Cyprus, a seventh century bishop of Neapolis. Quoted in K. Ware, *The Orthodox Way* (Crestwood NY: St. Vladimir's Seminary Press, 1995), 54.

relatively new city. Nevertheless, we can discern in this city an architectural tradition and urban planning, a cultural tradition, a tradition of freedom, a harmonious and ordered coexistence of its inhabitants. The opportunity is granted to each of the ethnic communities to be organized and developed based on their religious convictions and cultural particularities in the context and for the good of the Australian society. . . . This is why Orthodox Christians live comfortably within the tolerant and liberal Australian society, while also being enabled to contribute to it.

Coming from a rich tradition as ours, and encountering here the Mayor of the largest city in this free country, we feel that we share the same spirit. Therefore, we pray that the Lord will bless your efforts for the beautification of the city. May the development of the urban thread and of your city planning be rendered still more humane, for an application of modern technology that is beneficial to humanity and harmless to the environment, as well as for the creation of new spaces that will cultivate and promote the cultural activities of your citizens.

HUMANITY AND NATURE

The Ecumenical Patriarchate, too, has a particular interest in the preservation of an ecological balance. Creation is a gift from God to humanity, a companion and servant in the daily needs of humanity. It has been given to us in order to live in harmony with it, in order to use its resources with measure, in order to cultivate and preserve it, in accordance with the expression of the Old Testament: "Till and keep it" (Gen. 2.15). Within the unimpaired natural environment, humanity discovers spiritual peace and rest; and in humanity that is spiritually cultivated and possesses the grace of God and inner peace, nature recognizes its lord and companion. The Orthodox Church has always encouraged humanity to respect the works of God, while the Saints are considered the best friends of creation.

This is why we rejoice at being in Sydney and in observing the ecological balance here. You live in one of the most beautiful cities of the world in the midst of incomparable natural beauty. Your efforts toward a harmonious development of the urban thread with the surrounding natural environment are admirable. We pray that the city of Sydney will grow with the same respect that is due to creation and with the same attention that nature deserves, so that it may continue to be, even for those who come after us, a place of spiritual comfort and an opportunity to glorify the Creator.

Remarks at a Musical Reception, Sunday of Orthodoxy, Istanbul, Turkey, March 24, 2002

CREATIVITY AND CREATION

The Sunday of Orthodoxy appears at first sight to be a religious feast that exclusively concerns that segment of Christians who belong to the Orthodox Church. A more detailed examination, however, persuades us that it is essentially a universal feast during which truth—as the supreme desire and ambition of the human spirit—is honored. Naturally, the variety of human conceptions of truth is such, that we cannot say that we are approaching a unified and universal understanding of truth. Nevertheless, we are able to proclaim with conviction that the appreciation that all of us, and almost all people, have about truth is common, given, and established. In this respect, Christ's words, that "you shall know the truth and the truth shall set you free," are universally accepted, irrespective of the particular religious faith, knowledge, and philosophical stance of each person.

Even skeptics and all manner of agnostics, and in general all those who deny the human potential to approach and experience truth, cannot deny the Aristotelian assertion that humanity naturally thirsts for knowledge. They cannot deny the fact that such a natural attraction toward knowledge reflects or depicts a deeper and innate universal desire to achieve perfect and infallible knowledge of all things. This desire is what we otherwise call the knowledge of truth.

TRUTH AND BEAUTY

Behold, therefore, we are celebrating today a feast of Orthodox Christians, the feast of Orthodoxy.[3] It is a feast of true opinion, or rather of true knowledge. It is not simply a feast concerning the secular triumph of

---

3. The Sunday of Orthodoxy is celebrated on the first Sunday of Great Lent in remembrance of the triumph, restoration, and veneration of icons by decision of the Seventh Ecumenical Council in Constantinople (787).

one segment of people, who believe in certain truths. Rather, it is a universal feast, which emphasizes that any prevalence of a true opinion, or of a true knowledge, is universally profitable and worthy of more general celebration.

More particularly, from a narrow Christian perspective, the predominance of the opinion that the depiction of tangible reality is permissible today becomes a universal possession. For, the transmission of knowledge and the progress of science are not possible today without a form or picture or realistic representation, which the natural world assumes and through which we conceive the same natural world. Of course, we respect and even agree with the religious prohibitions against depicting the invisible Divinity. For, we do not permit imagination to substitute tangible perception. However, we are celebrating today the possibility of depicting this tangibly perceived reality. In any case, through such depiction, we are able to communicate with one another and to convey knowledge and sentiments to each other.

It is precisely this purpose that is fulfilled by today's musical reception. Through the universally accepted means of the musical harmony of notes and rhythms, the select composers and performers of the musical creations are conveying to us the truth that all of us share a potential for the common enjoyment of cosmic harmony. We can all enjoy and be grateful for this music. All people are able to delight in common. Therefore, the conception that the joy of wealth of one must correspond to the sorrow or deprivation of another is a delusion that is far removed from truth.

MUSIC AND BEAUTY

Naturally, the narrow-mindedness and greediness of so many people will not comprehend how participation in joy only increases this joy, while the sharing of sorrow only diminishes the sorrow. Yet, the present musical enjoyment confirms that the joy of one person among us can never constitute a hindrance in the joy of another. Indeed, our joint participation in this enjoyment extends the pleasure that we experience, inasmuch as it strikes chords deep within the heart. These chords produce sounds of interpersonal harmony and love, and which also reveal the deeper and archetypal truth that all people are created to coexist, to share in joy and gladness, to live in peace and harmonious coordination.

Consequently, the Feast of Orthodoxy reveals to us its universal significance as a celebration of truth. Furthermore, this feast celebrates the affirmation of creation as being very good, inviting us to an understanding of the world that shares in the overall divine harmony and is detached from any dissonance and lack of harmony that result from an extreme and partial emphasis of only certain isolated elements. The natural and environmental balance is necessary for the survival of the world's ecosystems; but it also leads to a moral balance between individualistic and communal ambitions, which are necessary for the smooth development of social systems.

The truth that sets us free indicates, through this music, that the harmony of our spiritual chords and rhythms is a necessary condition for the audience and delight of social, spiritual and universal harmony. It is for this harmony that we were created and intended. The Great Regulator of this universal harmony, according to whose rhythm the myriad of enormous galaxies as well as the boundless multitude of minuscule cells move, invites us to join in the harmony of the excellent unity of peace, love, and truth that it suggests. Let us obey in order to hear ineffable otherworldly words on earth, words heard only by the chosen that ascend to the heavens.

4

Ecology and Ethics

Virtues and Values, Responsibility and Justice

CONFERENCES AND SEMINARS

Message to the Inter-Orthodox Conference, Academy of Crete,
Greece, November 5, 1991

ECOLOGICAL RESPONSIBILITY

It is with much joy and deep satisfaction that we greet the inauguration
of this Inter-Orthodox Conference on the protection of the natural environment. The convocation of this conference was a particular desire and
hope of our late predecessor, Patriarch Dimitrios of blessed memory, who
convened it on the decision of the Holy and Sacred Synod, by means
of official Patriarchal Letters inviting their Beatitudes the leaders of the
Orthodox Churches to send representatives. The gathering of such representatives at the Orthodox Academy of Crete for this purpose surely already fills the blessed soul of the late Patriarch with joy. However, it also
moves us to offer praise to the Lord and express our warmest thanks to
all of you, who have gathered from near and afar either as delegates of the
Most Holy Orthodox Churches or as observers from other churches and
confessions and international organizations. We greatly appreciate your
favorable response to this initiative of the Ecumenical Patriarchate.

The importance of your conference hardly needs to be stressed. In his
message of September 1, 1989, the late Patriarch Dimitrios expressed the
deep anxiety of the whole of the Orthodox world and of every responsible

thinking person concerning the environmental disaster so rapidly overtaking us, for which our thoughtless abuse of God's material creation is entirely to blame.[1] Already "creation groans in travail together," in the expression of the Apostle Paul (Rom. 8.22), under the boot of human greed. The responsibility of spiritual and political leaders as well as of all people is enormous. Orthodoxy, too, must give her testimony. Your conference is invited to contribute in this respect.

We are present in spirit in your deliberations. Furthermore, we bestow on these deliberations our Paternal blessing, praying from our heart that the Holy Spirit will guide your work to full success so that through this means the holy name of the Triune God may be glorified, the bonds uniting the Holy Orthodox Church may be tied more firmly, and the salvation of humanity and the whole of creation may also be furthered.

*Opening Ceremony, Second Summer Seminar, Halki, Turkey,
June 12, 1995*

ENVIRONMENT AND ETHICS

Our Holy Great Church of Christ, the Mother Church, justifiably boasts in God for having hastened to be among the first to initiate a complete series of seminars on ecology within the context of her broader ministries in the world. The Church's efforts, of course, are not to be perceived as simply striding along with modern times. Rather, our concerns are rooted in the deeper conviction that by these initiatives the Church ministers thoroughly and accountably within the primary mission entrusted to it by God in history, namely the evangelization and salvation in Christ of humanity and the natural world.

Out of this deep-rooted conviction, we humbly believe that, during these distressing times in which contemporary humanity around the world finds itself, although it is by all means a noteworthy and honorable enterprise for one to speak casually of ecological concerns, there is often the danger of being misunderstood as someone going along with the "trend" of the times. Such a person may be misconceived as one who rather seeks and pursues what impresses instead of what actually illuminates and essentially contributes to the solutions of relevant problems.

---

1. See Chapter 1, "Call to Vigilance and Prayer."

Who can truly deny that, in spite of the honorable efforts made by various sectors in order to respond properly to the clamoring demand surrounding this issue, we all still continue to compete with each other on the problem of the environment, speaking and babbling rather than thinking and concurrently doing what is right?

This seminar, which with God's blessing is being hosted in this renowned Monastery of the Holy Trinity on Halki, on the theme "Environment and Ethics," is an attempt to restrain precisely this immense danger, which, as usual, silently and perhaps subconsciously threatens to foil much of the theoretical work done on this subject.

Consequently, as our paternal duty, we take this opportunity not simply to make these introductory remarks as though they were a formal greeting from the Mother Church, one in which we bestow our wholehearted Patriarchal blessing upon the seminar participants who have readily gathered here. Rather, purely out of pastoral concern, we wish to address a few thoughts on the topic at hand. In this way, we may perhaps from the onset place a finger on "the print of the nails" (Jn 20.25) in favor of a God-pleasing assessment of the obligations and duties of all men and women created in the image and likeness of God toward everything in the created universe within and around them.

A DIVINELY IMPOSED ETHOS

Therefore, permit us to state initially that the truly awesome endowment, which we so often voraciously lay claim to within nature, namely that we are created in the image and likeness of God, by definition predetermines an analogous ethos that is imposed upon us. Such an ethos is critical with respect to each other and ourselves as well as with respect to the microcosm and macrocosm around us. Only then can we truly satisfy God who created "out of nothing" everything that is "very good" (Gen. 1.31).

This means that, as the visible and living image of God in the world, humanity does not have the right to possess an *ethos* that is ungodly or unloving toward God. For, as a "partaker of the image but not as a custodian of it," the human person may, in fact, become the inciter and chief perpetrator of evil, which God as a fair judge providentially terminates upon our natural death "in order—as St. Athanasius the

Great[2] observes—that evil not be immortalized." Nevertheless, the ethos that springs forth from God and bears witness to the unapproachable and unknown essence of God is everywhere and always described as grace, throughout the Bible and God's revelation to us in general, as well as through the teachings of the God-bearing Fathers of the Church.

Speaking in the presence of Christian intellectuals, it would be meaningless and totally superfluous to clarify further that the most profound characteristic of the grace of the all-beneficent God is that it is totally free, that is to say, absolutely given and non-reciprocal. We should like to remind you, however, that, in the biblical account of creation, the grace of God is manifested initially as beneficence, goodwill, compassion, philanthropy, and the like. Yet, thereafter, following the fall and apostasy of humankind, it is revealed as mercy, forbearance, expiation, restoration, reinstatement, and adoption.

DIVINE ECONOMY AND HUMAN ECOLOGY

Nevertheless, we should unequivocally state that, in both instances, the divine will of God was always manifested in the form of law and order, which no one has the right to violate without being punished because the entire design of the all-beneficent Creator constitutes a unique and indissoluble divine "economy." Thus, whether we speak of natural, moral, or spiritual principles, we acknowledge and emphasize the same infinite grace of God, confident that "divine economy" is always the solid support of ecology as a whole.

All these things are preeminently indicative of the compassion of an omniscient and omnipotent God. And they should certainly and ceaselessly constitute a guiding policy of our own *ethos* in the world. For if God, according to Plato, "were perpetually to geometrize [lit., to be the measure of everything in the created world]," then it follows that humankind must always read and obey all the laws pertaining to the world.

Within the framework, then, of this pious awareness concerning the world, we should seriously bear in mind also the rudimentary sequence within the entire order of creation in which God in the six days of

---

2. St. Athanasius (293–373) was a deacon during the First Ecumenical Council of Nicea (325) and later served as Archbishop of Alexandria.

creation classified everything, as it is known, from the less to the more perfect. Recognizing and acknowledging such an inverse hierarchy, perhaps then we shall respect anew the unquestionable and mystical sanctity possessed by the material world created before humanity not only for itself as a work of God "created out of nothing" (*ex nihilo*) but for the "being" and the "well-being" of humankind. Today, we are not being very eloquent when, for the sake of brevity, we characterize the created material world as the "natural environment."

We wish to say a word, in closing, on the famous scholastic axiom of the Western Church *gratia praesupponit naturam* (i.e., "grace presupposes nature"),[3] upon which essentially the whole of the West built both its past and present theories concerning Natural Law as well most of its sociopolitical concerns. This principle will always fall short and suffer injustice as a result of its one-sidedness, especially in regard to fallen creation in general, so long as it is not completed by the supplementary correction *natura praesupponit gratiam* (i.e., "nature presupposes grace"). This corrective balance was appropriately suggested in a relevant study made some time ago by a hierarch of our Ecumenical Throne. For, indeed, it is only within the liturgical conscience of Orthodoxy that the whole of God's creation is illumined and redeemed by divine love.

Opening Address, Fourth Summer Seminar, Halki, Turkey, June 25, 1997

ENVIRONMENT AND JUSTICE

It is a well-established fact that care for the environment of our human family regarding ecological matters constitutes a most urgent question for each and every human person. With every passing day, the danger threatening the survival of life on this beautiful planet of our universe proves to be yet clearer and ever present. Today, a host of international organizations, governments, and leading non-governmental bodies are sending the same message, in no uncertain terms, that they are bearing down on the very visible danger that is posed by the real disturbance in the ecological balance. Each has set forth proposals for the prevention of the certain destruction of our planet of which we have been forewarned.

---

3. See, for example, Thomas Aquinas, *Summa Theologiae* I–II, 98–105.

For this reason, the Ecumenical Patriarchate, the Mother Church, has taken this initiative and joined its own voice to those of many others and has taken diligent and incisive action for the protection of the environment, inasmuch as our material world is first and foremost a spiritual organism. The Church is compelled first by its love in Christ for our endangered fellow human beings, and also by its responsibility not only to teach the faith, but to practice it (James 2.14–17).

So it is that, as in many other matters of faith, the Church risks being condemned as indifferent unless it speaks out appropriately with the "word of truth." Failure to do so would result in opening the field to other voices, and it could sometimes result in the consequence that we, the believers in Christ, might not even recognize what our own faith teaches about these issues.

Therefore, our actions are out of love and a sense of responsibility in order that we may speak the truth and be of service to our fellow human beings. Our contribution to these efforts is the annual organizing of these environmental seminars here in the sacred monastery of the All-Holy Trinity at Halki. The theme of this year's seminar is "Environment and Justice."

At first glance, these concepts might appear unrelated, but they are most certainly worth a closer look. In Holy Scripture, justice does not have only or principally the current meaning of the dispensation of justice, namely of justice being served. Justice carries a more extensive and comprehensive sense of virtue, such as expressed in the well-known aphorism of Aristotle: "Every virtue is contained in justice." The just person does more than merely comply with the law; the just person bears within oneself a higher conception of justice, namely of the perfect relationship of all things to one another. Thus, the just person is virtuous in every respect. For example, Holy Scripture describes as just whosoever shows mercy all the daylong and freely gives of oneself, without any legal obligation. Likewise, Scripture portrays St. Joseph the Betrothed as being a just man, namely as being virtuous in all things, precisely because at the moment that he thought the all-holy Virgin was guilty of illicit union (the result of discovering her pregnancy), he did not wish to make an example of her or subject her to the punishment prescribed by the law, which was death by stoning.

If justice is identified with this correct understanding, then it becomes immediately apparent that the contemporary acute ecological problem has its root precisely in the lack of justice, in the lack of that comprehensive virtue of possessing all virtues.

REPENTANCE, RECONCILIATION, AND RESTORATION

Humanity freely departed from this virtue of justice when, at the prompting of the serpent, Adam declared his autonomy from God and thereby overturned the relationship of love and trust toward God. As a result, we came to sense our own vulnerability; we began to know remorse, to hear the approaching footsteps of death, and since that time we have been filled with fear and anxiety (Gen. 2.9–11). Recall how when Adam was called by God to account for his disobedience. Instead of humbling himself and seeking forgiveness, he placed the blame on God Himself, claiming that Eve was the cause, that helper whom God had provided for him. Thus, Adam lost the opportunity for repentance and for the restoration of love that would have absolved him from the fear of living the remaining years of his life in the inherited fear, which has endured in every generation since.

From that time, humankind has been engaged in the Titanic, if not Sisyphean, attempt to reclaim the power of kingship over creation through its own efforts, which were forfeited by this disobedience. Humankind arrogantly struggles through its own means to establish itself as God, to acquire for itself the very powers of divinity. Quite obviously, we have not succeeded in obtaining our desire. Surely there are many and varied passions of humankind that impel us to differing actions. Yet, beneath them all, one discovers our basic fear and fundamental uncertainty about the future. Consequently, we struggle to find substitutes for real hope and reassurance by which we can achieve security. Unfortunately, these efforts fail to establish us on the only true and firm foundation that is Christ, for only trust and love for Him can cast our fear.

Thus, we see humanity striving to shore itself up by accumulating wealth, and by manufacturing weapons to see revenge, and often to destroy one's fellow human beings as an enemy, real or imagined. We contradict ourselves by exterminating certain creatures as being bothersome while affirming that these same beings protect us from ones that are even more dangerous. Humanity has not only sought to take from the natural world that which is necessary for its own stability and survival, but often seeks to satisfy its perceived and ultimately false psychological needs, such as the need for self-display, luxuries, and the like. Twenty percent of humanity consumes eighty percent of the world's wealth and is accountable for an equal percentage of the world's ecological catastrophes. One cannot characterize this situation as just; and what is more, this injustice has had a direct impact on the ecology of the environment. However, it is plain that this numerical minority of financially powerful people is not the only cause for the ecological ruin of our planet. Every person ruled by instinctual fears attempts to exploit and loot nature. Consider the willful scorching of the earth, over-fishing, wasteful hunting, excessive and dangerous recycling of resources, and other similar "injustices" against the ways of nature, which share in the responsibility for this ecological spiraling down.

During the course of this seminar, you will have the opportunity to study these issues in detail. Our Modesty desires to convey to all of you participating here our heartfelt welcome, together with our paternal prayer that you might worthily actualize the possibilities available to you during the course of this seminar. We encourage you to exhaust the present theme, which we have raised here in a few points, just to stimulate your minds, by engaging your work from every perspective and by gleaning ideas and hypotheses, which are useful for all of humanity. It is indeed just for each of us to shoulder the burden of his or her own responsibility and not to remain silent, thinking that others will take charge and be liable, and that we can remain an isolated, insignificant factor. Only if there is "just" comprehension by each of us of our responsibility to work together—all of us—in a universal effort, will we be able to hope for a better world.

Foreword to Proceedings of the Fourth Summer Seminar, *June 2000*

JUSTICE—ENVIRONMENTAL AND HUMAN

For the Orthodox Church, the worldwide ecological crisis constitutes the expression of a more comprehensive crisis of the human race and concerns all of the critical issues faced by humanity. The crisis faced by creation cannot be examined exclusively from the perspective of the numerous and undoubtedly beneficial branches of Science and Technology. In order to secure long-term solutions and avoid repetition of errors of the past, we are obliged to study the ecological crisis, too, in light of the more general moral crisis.

One of the more fundamental problems, which constitute the basis of the ecological crisis, is the lack of justice prevailing in our world. By "justice," we mean not only the legal correspondence of giving and receiving, of transgressions and consequences, of offering and reward, but the more inclusive virtue that lies beyond the narrow fulfillment of obligation. The liturgical and patristic tradition of the Church considers as just, that person who is compassionate and uses love as one's sole criterion. Justice extends even beyond one's fellow human being to the entire creation. The burning of forests, the criminal exploitation of natural resources, the gap between the wealthy "north" and the needy "south," all these constitute expressions of transgressing the virtue of justice.

With the ultimate goal of studying the ecological crisis comprehensively from the viewpoint of justice, the fourth annual Summer Ecological Seminar was held June 25–30, 1997, in the Holy Patriarchal and Stavropegic Monastery of the Holy Trinity at Halki, with the general title "Environment and Justice." It is clear from the proceedings that this seminar succeeded in an inter-disciplinary approach of the subject. Participants also included representatives of the Orthodox Church, other Christian confessions, and other religions; governmental authorities; and environmental, scientific, and forensic experts, all of whom provided invaluable resources for those who are interested in just solutions to the ecological problem that challenges humanity today. May the grace of the Creator of all the creation, our Lord and God and Savior Jesus Christ be with you all.

Opening Ceremony, Halki Ecological Institute, June 13, 1999

PRINCIPLES IN PRACTICE

The beautiful Black Sea, around which so many renowned ancient civilizations flourished, constituted a very useful channel of communication between peoples and cultures throughout the centuries. Even more so, it is also a biotope, a living habitat, which for thousands of years existed under balanced conditions that allowed for the preservation of life both under the waters and on the region's surrounding shores; a life which was nourished by its natural recycling and which nourished its inhabitants as well.

The linkage of the Black Sea with the Mediterranean via the Bosphorus also allowed the Mediterranean peoples to travel by sea to the North and, conversely, for those in the North to sail south. Even though there have been many military conflicts in this region, today we see that this mode of communication contributed to the cultural development of neighboring peoples. By means of the mighty flowing rivers, in particular the Danube, the Black Sea allowed communication among the peoples living inland, thus constituting a center of communication, the significance of which continues even until today. The fact that many rivers of Europe, Asia, and Asia Minor flow into the Black Sea make it the recipient of the refuse settlement that these rivers deposit from sources that surround them and even those that are remote.

Unfortunately, the description by Herodotus about the primeval belief in the sacredness of rivers that existed among certain peoples living by them—a belief imposed upon them for reasons of religious conviction, that they must not pollute the rivers—is not a generally accepted credo of humankind. To this we must also add rational behavior, which refuses to respect the deeper and truer justification of such a principle and belief. The denial of that which is not rational, which in turn leads one to "reason" as the only criterion of truth, and having denied the pedagogical power of the myth with its higher rationality found hidden in its message, has led modern man to a shortsighted, selfish, and pettily opportunistic state. We experience the consequence of this state as a foretaste of our biological death, which unexpectedly comes upon us and for which we prepare ourselves by our own so-called rational energies, which are, in fact, foolish and irrational.

The overproduction and over-consumption of toxic substances in both industry and agriculture—not to mention war—directs those substances that are non-degradable toward an inert and harmless body of rainwater intended to cleanse the surface of the earth and for irrigation, into the recipients of the flowing waters of the rivers, that is, into the oceans. The Black Sea in this case, being relatively small in size, is the recipient of a disproportionate quantity of pollutants. As a result, it is constantly and intensely being overburdened even more so than are the great Pacific and Atlantic Oceans. Indeed, this apparently is because of the industrialization, the over consumption, and the overall changes and condition of life of modern society.

GENERAL CAUSES AND ROOT CAUSE

Even if these reasons, indeed, contribute to the pollution of the Black Sea, we insist on characterizing them as "apparent causes," because we consider the "true cause" to be the destruction of that which is religious piety within the human heart over and against the evil that can be carried by the rivers.

This piety that the ancient peoples had, referred to by Herodotus, elevated them to a level of spiritual civilization higher than our own inasmuch as we do not have the delicateness of feelings and the sensitivities of our responsibility in facing our fellow human beings. Thus, this entrenches us behind the egotistical pretension that our higher logic does not allow us to accept the sacredness of a natural thing such as a river. In this, we prove ourselves to be foolish and incapable of thinking on a plane higher than our small-minded reason in order to understand that the sacredness of rivers and of all creation exists and is a given, just as is the sacredness of the human person, which nature, itself, is ordained to serve. All that was created "good" by the All-Good Creator participates in the sacredness of its Creator. Conversely, disrespect toward it is disrespect toward the Creator inasmuch as the arrogant destruction of a work of art is an insult to the artist who created it.

Consequently, if we desire to improve the situation, we must restore in the hearts of the members of our society the sensitivity that was held in the hearts of our ancestors, whom Herodotus mentions. In other words, we must restore respect to the truly existing sanctity of life, which is in peril because of our shortsighted and egotistical polluting actions.

The riparian people are so numerous and so greatly dispersed, even among many nations, that there is no possibility of fully monitoring them. The successful method of avoiding river pollution, a method discovered by the early ancestors of Herodotus, not as a useful lie but as a most profound truth, is denied and rejected by some of us modern demythologizers and intellectualists as confusing the supernatural with the one-dimensional nature of the world, according to our deluded perception. The result is that we have allowed human individualism to act in a very shortsighted way, inasmuch as the transferal of pollution far away from us satisfies us, and we feel secure. Yet, we do not consider the fact that we are thus setting into motion a vicious circle of a mutual transferal of pollution and a vain struggle of repression and healing, where in fact only prevention can save the situation.

Such prevention can be achieved only if all members of society regard it as their moral and above all their religious obligation not to cast their wastes upon other fellow human beings. This is understood and socially accepted as an obligation of life. Yet it must be expanded also to our broader economic and biomechanical moral conduct.

THE WONDER OF CREATION

We are therefore obliged to recall the concept and the ontological acceptance of the sacred in our daily life, and especially wherever we would not otherwise recognize its place, namely in our commercial and professional activity. This is also the reason why the Ecumenical Patriarchate, which has a purely religious mission, is mobilizing and sometimes initiating efforts for the protection of the environment, which at first glance seem to refer only to the material world.

However, this superficial evaluation is not true, because on the one hand the protection of the environment is not for us a mere worship of nature; it is not the adoration of creation, but the veneration of the Creator. On the other hand, it is an invitation for all of us to accept the sacred and the holy in our lives.

In this sense, all of our theologians and clergy, who have recently admired the beauty and diversity of our natural environment, must become conscious of the fact and convey the message that the life of our faithful should not be exhausted in individual morality, but should be extended out of love to the avoidance of those long-term consequences that may harm our fellow human beings who live hundreds of miles from us. Our Lord has taught us that our righteousness must exceed the righteousness of the scribes and the Pharisees, or, at least that of those whom Herodotus referred to as the ancient Persians, who neither spat nor washed their hands in the river, and who also encouraged others to do likewise, out of great respect for the rivers.[4]

We are certain that the Halki Ecological Institute will offer you the opportunity and the motivation to study more deeply not only the technical and humanitarian parameters of the problem of the pollution of the Black Sea and every other natural environment, but also its Christian and theological perspective. This will enable you to become interested and more deeply conscious of your mission to work with love and piety, and to cooperate with each person dealing with the subject of the environment, for its protection from actions that create pollution and destroy the environment. This sensitization of ourselves and of those around us, especially those who direct the great pollutants, together with the voluntary avoidance of ecologically destructive lifestyles by members of our society, and their influence over those who do not accordingly conform, constitutes the most fruitful way of environmental correction and of revival for the Black Sea and every burdened ecosystem.

We repeat once again from this position our invitation to all of you—to the Orthodox and other churches, and to the religious leaders of the faiths in the neighboring region, as well as in the depths of Europe, Asia, and Asia Minor, from where rivers transfer pollution and especially toxic wastes into the Black Sea—to convey to all peoples the need to raise their awareness about such pollution, to the level at least of those people who some 2,500 years ago would not even wash their hands in the rivers.

We thank all those who are mobilized together with us for the reintroduction of the sense of sacredness as the guideline for our life, as well as all those who from whatever position carry on the struggle for the preservation of life in the Black Sea, thereby contributing to and assisting our neighboring peoples.

Message to the International Conference on Ethics, Religion, and Environment, University of Oregon, April 5, 2009

AN ALTERNATIVE ETHIC

It is with great joy that we convey our wholehearted greetings to the organizers and participants of the conference Ethics, Religion, and the

---

4. *Kleio*, par. 138.

Environment held at the University of Oregon on the occasion of the meeting of Chairs from throughout the world in the UNESCO program of Intercultural Dialogue for Inter-Religious Understanding. Unfortunately, we shall be unable to be with you in person, but we shall most surely be with you in spirit and prayer inasmuch as the themes of your conference lie so close to our heart and to our ministry.

Some fifteen years ago, the Ecumenical Patriarchate was privileged to convene—in association with HRH Prince Philip, Duke of Edinburgh, President of the Worldwide Fund for Nature—an international educational seminar at the Theological School of Halki on Environment and Ethics, drawing attention to the unique responsibility that humankind has toward the preservation of our natural resources. Indeed, through our ongoing interfaith dialogues, we have insisted that the survival of our planet provides a fundamental basis for mutual understanding and tolerance among adherents of the world's religions.

Therefore, it is with a deep sense of spiritual satisfaction that we learn of your unique initiatives in the academic world, through the UNESCO Chair in Trans-Cultural Studies, Inter-Religious Dialogue and Peace, within a State already deeply conscious of our impact as human beings on the natural environment. The University of Oregon, as well as the region in general of the northwestern United States, has a strong tradition of concern for protecting and preserving the environment.

We are convinced that the root cause of all our difficulties consists in human selfishness and human sin. What is asked of us is not greater technological skill but deeper repentance, *metanoia*, in the literal sense of the Greek word, which signifies fervent "change of mind" and radical transformation of lifestyle. The root cause of our environmental sin lies in our self-centeredness and in the mistaken order of values, which we inherit and accept without any critical evaluation. We need a new way of thinking about our own selves, about our relationship with the world and with God. Without this revolutionary "change of mind," all our conservation projects, however well intentioned, will remain ultimately ineffective. For, otherwise, we shall be dealing only with the symptoms, not with their cause. Lectures and international conferences will invariably help to awaken our conscience, but what is truly required is a baptism of tears and, ultimately, a sacrifice of means.

It should be noted here that, just as environmental ethics are a fundamental religious and not merely a basic secular issue, so too sacrifice is primarily a spiritual issue and less an economic one. In speaking about sacrifice, we are talking about a matter that is not technological but ethical in nature. Indeed, environmental ethics is specifically a central theme of your symposium. And so it is crucial for us to remember that, while we often refer to an environmental crisis, the real crisis lies not in the environment itself—which suffers the consequences of our actions—but in the human heart. The fundamental problem is to be found not outside but inside ourselves, not in the ecosystem but in the way we think about, perceive, and treat this ecosystem.

Moreover, respect for creation stems from respect for human life and dignity. It is on the basis of our recognition that our world is created by God, that we can discern an objective moral order within which to articulate a code of environmental ethics. In this perspective, believers of all religions have a specific role to play in proclaiming moral values and in educating people in ecological awareness, which is none other than responsibility toward self, toward others, toward creation.

We are urging for a different and, we believe, a more satisfactory ecological ethic. This ethic is shared with many of the religious traditions represented at your conference. All of us hold the earth to be the creation of God, where God placed the newly created human "in the Garden of Eden to cultivate it and to guard it" (Gen. 2.15). God imposed on humanity a stewardship role in relationship to the earth. How we treat the earth and all of creation defines the relationship that each of us has with God. It is also a barometer of how we view one another. For, if we truly value a person, we are careful as to our behavior toward that person.

It is with that understanding that we endorse your extraordinary efforts in organizing this international and interfaith conference to contribute to the great debate about the relationship between ethics, religion, and environment. We must—all of us, each from our own perspective and discipline—become spokespersons for an ecological ethic that reminds the world that it is not ours to use for our own convenience. It is God's gift of love to us and we must return this love by protecting it and all that is in it for the sake of future generations. May God bless your gathering and inspire your deliberations.

Foreword to Greece: Land of Diversity, *August 26, 2010*

DIVERSITY AND BIODIVERSITY

If the environmental crisis is the greatest threat for human survival and welfare in the world's historical journey, it is evident that every initiative that contributes to the resolution of the ecological problem, to the development of ecological awareness and responsibility, as well as the establishment of an ecological civilization is crucial, invaluable, and praiseworthy.

Unfortunately, today, despite the crisis that we face, the contemporary self-ordained man-god continues—under the ideological flag of progress or economic development and with the instrument of technology—to ignore limits and measures in destroying the conditions of life on this planet. At the outset of the twenty-first century, we continue to pillage the land, to destroy the environment, and to reduce biodiversity with precipitous acceleration. We remain blind to the undeniable truth that there can be no sustainable development at the expense of the natural environment. The irrational and arrogant objectification and exploitation of nature seem to lead to inevitable ecological devastation.

What is demanded is vigilance and action! The resistance of those who bear and express the ecological civilization has critical significance for the common future of humanity and the natural environment. It is insufficient to be merely stressed about nature; what we need is to struggle for its protection. The rapid loss of biodiversity worldwide renders our common action mandatory for its survival.

The Orthodox Church reminds and underlines the cosmological dimension of sin, the destructive environmental and social consequences of the spiritual crisis that confronts contemporary humanity, and invites us to become engaged with the crisis in moral values. The proposal of life conveyed by Orthodoxy is a eucharistic relationship with nature, a human being as the "priest of creation," as its protector and steward, who cultivates and cares for it, beautifying and returning it to God in a spirit of humility.

It is our hope that the splendid volume *Greece: Land of Diversity*— promoting the natural treasure of Greece, informing about the wonderful array of flora and fauna, while inspiring and energizing for their protection—inasmuch as it also propitiously appears during the "International Year of Biodiversity 2010," will contribute to an ecological awareness, a change in attitudes, and the proper vigilance on the part of all before the imperiled natural environment. With its numerous and diverse ecosystems and biotopes, Greece needs eco-friendly citizens. It requires coordinated action to deter the imminent dangers for its "very good"—to adopt the biblical phrase—natural environment.

We conclude this foreword with the prayer that the festive volume *Greece: Land of Diversity* will be duly appreciated and your overall endeavors may prove fruitful. We congratulate you, we praise the ecological initiatives and contribution of the Greek Society for the Protection of Nature, and we express our esteem toward the Bodosaki Institute for covering the costs of this publication.

Therefore, we convey to all of you our paternal and Patriarchal blessing, invoking upon you the grace and boundless mercy of God.

Religion, Science, and the Environment Symposia

Closing Ceremony, Symposium I, Patmos, Greece, September 25, 1995

CALL TO ACTION

At this moment, we are receiving from your blessed hands the results of the inestimable collected effort of the God-loving participants and speakers of this symposium. And we are deeply moved by the same emotion, which permeates the liturgical celebrant of the Most High when he receives from the hands of the faithful who approach the church in order to offer the bread and wine, which are to be consecrated in order that their offerings might constitute immortal nourishment for the whole world.

Therefore, along with their Beatitudes and their Eminences the Presiding Hierarchs and your honorable persons, we hereby also bless these joyous fruits of your symposium, which by God's grace has drawn to a close. We express our congratulations and gratitude to you all for what has been achieved. We are certain that, from this point forward, your scientific and intellectual vigilance will remain the unswerving compass by which you will guide your colleagues and students to more illumined horizons.

During this official occasion, we wish to assure you, on behalf of the Church, that, aware as we are of the significance of your contribution of empirical work in all areas of Godly inspired thought and science to the progress and welfare of the world, we shall not cease to pray for your good health and increasing achievement.

In closing, we wish to add one simple observation, which is already known to everyone, namely that the destructive deterioration of the environment is taking on multiple and threatening dimensions. Therefore, we must not be content with verbal protests, but we instead must proceed to continually stronger and more effective actions, each from its own part and position. For, pollution is dangerously spreading and rapidly increasing. Indeed, quite possibly and, God forbid, according to the calculations of the experts, quite probably, pollution will become impossible to control. *We cannot remain idle.*

May the enlightenment of the Paraclete always shine in your steps and in your actions within the course of your research and science, for your own benefit and for that of all your fellow human beings and the whole natural world.

Official Banquet, Symposium II, Batumi, Georgia,
September 21, 1997

ECOLOGICAL CRISIS

The contemporary ecological crisis in its diverse forms has been characterized and indeed is the most immediate threat not only for the overthrowing of the functional balance of our ecosystem, but also for the continuation of life on our planet. The hole in the ozone layer, the greenhouse effect, the pollution of the atmosphere, the defilement of the terrestrial and underground water-table, and in general all of the principal and consequent phenomena, which, in a particularly intense manner bear upon the contemporary ecological problem in every concrete place, have already been pointed out at the international and local level by the appropriate international organizations and by informed, sensitive local representatives, both governmental and other.

The impressive presence and the warm reception accorded to the international ecological symposium by the political, religious, and intellectual leaders of the lovely city of Batumi, give expression to the agony, common to us all, concerning an ongoing acute problem, the universality of which is lived out with its painful consequences in every particular place and by every particular people. The impressive response, however, of the people to this initiative of the international ecological symposium confirms not only the wider consciousness of the tragic dimensions of the environmental crisis, but also the people's wholehearted disposition to contribute in every useful way toward the protection of their environment.

The beautiful city of Batumi, with its surrounding region, was always a major crossroads of peoples, religions, and cultures. Commerce, in previous times, and thriving industry in more recent years, were influenced to a great extent by the city's natural link with the Black Sea, which was indeed a "Hospitable Sea" for all the people of the North and the South, the East and the West. Today, the beautiful coastline of the surrounding region of the city, which becomes crowded with thousands of tourists during the summer months, is in danger of losing its attraction because of the dangerous pollution of the Black Sea, into which radioactive and other harmful waste is emptied not only from countries surrounding the Black Sea coast, but even from those further afield.

The Black Sea, with the uncontrolled and continuous outlet of radioactive and other kinds of waste from European countries, is in danger of being labeled as the most polluted sea in Europe. This has particularly dangerous consequences, not only for the peoples of the entire Black Sea area, but for all the Eastern Mediterranean, unless the necessary measures are taken in due time for its vital protection. The uncontrolled influx of radioactive and destructive waste-matter, primarily from countries to the north and the west, with principal channels the major rivers, just as the impossibility of their disposal through the very small and shallow outlets of the waters of the Black Sea, have rendered the deeper sections of its sea bed a dangerous depot of the waste of the countries of Central and Eastern Europe.

CRY OF AGONY

The theme of this international ecological symposium is a responsible cry of agony for the visible danger of total catastrophe for the Black Sea, that natural source of life for all the peoples of the Black Sea region. The undertaking of the symposium by ship, with scientific reports presented by eminent scientific personalities, combines, on the one hand, the living link with the problem and the necessary proposals for its solution, and on the other hand, the contact with the appropriate political and intellectual leaders of the Black Sea region, who live out the painful consequences of the continually aggravated environmental crisis and who are willing to join us in the attempt to overcome the threat or disaster.

The formulation of the main theme of the symposium is in the form of a question. Obviously, this is not because the eventual danger is uncertain, but because our cooperation for the aversion of the danger must be constant and must be proclaimed in every quarter as a cry of agony for the future of the peoples of the Black Sea region. In this common effort, science will offer the correct processing of the necessary proposals for the control of the influence of radioactive and other polluted waste; religion will enhance the common conscience for every believer of our personal responsibility before God for guarding the integrity of divine creation; and the political leadership of the peoples will support, through every useful means, the protection of the natural resources of life for all the peoples of this region.

Scientific research observes that the Black Sea is in danger, political leadership becomes receptive to the messages of the dangerous threat, and the peoples live out these tragic consequences in their everyday life. The reversal of this crisis is undoubtedly difficult because it has become too much involved with the vicious circle of a consumer interpretation of the link between humanity and the ecological environment, but it is more than ever necessary, even if many years of hard struggle are needed. The cooperation of religious leaders with all the other factors and all the deployable means at their disposal, for the effective confrontation of this serious problem, will give to concrete proposals the dynamism of the faith of the peoples involved for the protection of the ecosystem from the plundering mania of contemporary humanity.

Arrival at Novorossisk, Russia, Symposium II, September 22, 1997

PROMPT AND PROPER RESPONSE

An international symposium, in which distinguished representatives of religion, science, and the competent organizations for the protection of the environment participate, could be characterized as an opportune initiative for greater conscientiousness of the dangerous dimensions of the modern ecological problem. However, this international ecological symposium, which has as its main topic the recognized danger of the destruction of the Black Sea, assumes special importance because it analyzes through well-founded reports the destructive factors of the specific problem and proposes specific solutions for its timely confrontation. At this symposium, the theoretical declarations of international organizations dealing with the seriousness of the modern ecological crisis come into contact with the specific dimensions of the problems of the Black Sea, the correct confrontation of which calls for the immediate move from theoretical accounts to practical proposals for the prevention or cure of these problems.

The visit to this splendid city of Yalta offers a representative picture of the exceptional importance of the Black Sea for the life of the peoples of the area around it, while at the same time making more tangible the threat of the results of this proven ecological crisis for subsequent generations. It is now a common position in science that the correct confrontation of whatever problem presupposes the correct description of all the factors that bring it about. The proposals of the specialist speakers at the symposium are describing with characteristic completeness the causes of the problem and submitting well-founded proposals for its timely and effective confrontation.

Nevertheless, in the final analysis, all the disturbing descriptions of the causes that relegate the Black Sea to a storehouse of radioactive and other waste materials of the countries of Central and Eastern Europe, as well as all the proposals for the confrontation of the problem, hold man himself as their common denominator. This is the tragic reality of the problem. The arrogant apostasy of humanity from the deeper reason of its relationship with our divine Creator's creation incites the presumptuous and improper exploitation of the ecological environment. However, the painful consequences of this apostasy for human life itself and for life on our planet, as is now indicated by the anxious declarations of all the responsible international organizations, render essential the restoration of humankind's inner spiritual equilibrium so that we may become more fully aware of the threat.

PASSIVE OBSERVERS OR ACTIVE PARTICIPANTS?

The visible danger of the destruction of the Black Sea, as this has been described with tragic clarity in the addresses of the specialist speakers and in the interventions of the distinguished members of the symposium, constitutes what is perhaps the most characteristic confirmation that humankind, with its arrogant preferences for the violent submission of its natural environment, has eventually become a timid and ineffective observer of the tragic results of its choices. The purpose of our symposium is to evaluate the conclusions of science and the principles of religion for the restoration of the genuine relationship of humanity with the natural environment, in order to assume personal responsibility for the specific problem of the Black Sea and more generally the problems of our planet.

The warm reception accorded to our symposium by the civil and intellectual leadership of the beautiful city of Yalta constitutes not only a significant encouragement for our initiative, but also an expression of active participation in the struggle to save the Black Sea. If all the cities of this region support the final proposals of the symposium with consistency and continuity, then the hopes of us all will not be disappointed, and the Black Sea will become a credible commendation for confronting similar ecological crises.

Official Banquet, Symposium II, Constanza, Romania, September 26, 1997

WORSE THAN THE PLAGUES OF PHARAOH

One of the most urgent problems of our era is that of the environment, given that for the first time in the history of humanity, nature is threatened together with humanity in a definitive and irreversible manner. The new element, which elevates the ecological problem to a level in a way above that of the plagues of Pharaoh, is the irreversible character of many of the catastrophes that occur. The ills afflicting the environment are difficult to correct, and for that reason it is essential for us to take preventative measures.

Our common responsibility for nature constitutes the beginning of every sensitivity with reference to ecological matters. Believers and unbelievers, religious people and atheists, regardless of sex or nationality, social position, economic situation, political standpoint and position on world-theory, are equally affected by the destruction of nature, which is classed under the heading of the environmental question. For this reason, precisely, all of us without exception fully bear the responsibility for any indifference in the face of the ecological threat. All of us are affected by the destruction of the natural environment and, without exception, we should all assume our responsibilities and undertake initiatives. All of those who violate nature are perpetrators today, and tomorrow become victims of their own violence.

RESPONSE TO EARTH—RETURN TO GOD

In participating in the effort to protect nature, the Ecumenical Patriarchate does not go so far as to idolize it, but tries to make conscious responses so as to conform to the divine commandment to labor and to preserve the environment in which God placed humanity. One criterion of environmental behavior for us is the liturgical ethos of the Orthodox tradition. This ethos is summarized, as it were, in the eucharistic proclamation: "In offering to Thee, Thine own [gifts] from Thine own, in all and through all—we praise Thee, we bless Thee, and we give thanks to Thee, O God." We do not despise God's gifts, among which is the "very good" nature. Rather, we offer this gift back to God with awe and due respect.

"Gift" (*doron*, in Greek) and "gift-in-return" (*antidoron*, in Greek) are the terms by which it is possible to signify our Orthodox theological view of the environmental question in a concise and clear manner. Nature is the *doron* of the Triune God to humankind. The *antidoron* of humankind to its Maker and Father is the respect of this gift, the preservation of creation, as well as its fruitful and careful use. The believer is called upon to celebrate his or her daily life eucharistically, that is to say to live and to practice daily what he or she confesses and proclaims at each Divine Liturgy: "In offering to Thee, Thine own [gifts] from Thine own, in all and through all—we praise Thee, we bless Thee, and we give thanks to Thee, O God." *Doron* and *antidoron* exemplify the realized eucharistic conduct of the Orthodox Christian, which is prefigured in an exemplary manner in the liturgical practice.

Address at Parliament House, Symposium III, Bucharest, Romania, October 24, 1999

BYZANTINE HERITAGE

Having entered this very beautiful hall of the House of Parliament of the representatives of the glorious and pious people of Romania, we are overwhelmed by the weight of a historical memory that unfolds before us like a vision of the indissoluble spiritual bonds between the ecclesiastical and spiritual leadership of the much loved Romanian people not only during times of fortune but also in times of trial. In the brightness of your faces we see reflected in an exceptional way the communion of faith in the bond of love, which springs from our common experience of all the Orthodox peoples.

The Holy and Great Church of Christ of Constantinople and all the local Orthodox Churches have a common reference to the spiritual heritage of Byzantium. This heritage has stood for the transcendence of the anthropocentric philosophy of classical Greek antiquity by accepting the anthropocentric teaching of Christianity. Such transcendence did not imply the rejection of all the achievements of ancient Greek thought, but the selective acceptance and application of all good elements therein. We see this selection imprinted in a superb manner in the literary work and theology of the great Church Fathers. It has to do with their employment of the terminology and methodology of the Greeks that were imbued with Christian concepts and aims. Thus, the Byzantine Orthodox Church was able to express successfully by means of the advanced Greek language and the elaborate Greek logic the Christian principles and convictions concerning the God-Man, God, and the world, which were new and revolutionary to the Greek world. At the same time, it contributed auspiciously to the spiritual life of the Orthodox Church as a whole, by excluding all Judaizing and Hellenizing heresies and remaining firmly attached to the Christian truth of the Cross, and of the salvation that springs from it. It did this although this truth was foolishness to those possessed purely of the Greek spirit and a scandal to those animated with pure Judaic spirituality (1 Cor. 1.23). Thus, the Mother Church of Byzantium formulated the specifically Orthodox proposition concerning our liturgical relation to God and to the world. It is this formulation that inspires even today the proposition of Orthodoxy concerning the proper approach to the contemporary problems of humanity and the world.

LITURGY AND LIFE

It is the common consciousness of all Orthodox peoples throughout the world that the spiritual heritage of Orthodoxy impregnated over the centuries their public and spiritual life, defining in an impressive way the particular ethos of the Orthodox peoples. This was done not only with respect to the liturgical appropriation of the experience of the faith, but also with respect to the extension of this experience to the secular realm. In the Orthodox tradition and spirituality, the world is initially very good but subsequently becomes a rebellious creation, within which humanity is called to achieve through divine grace, and personal willingness and endeavor, the assimilation to God and deification by grace. Through the Orthodox Church, the sanctifying and restoring divine grace of God is extended to the entire cosmos. This is the grace that springs from the Holy Altar, on which the mystery of the divine economy in Christ is constantly celebrated, and the sacredness of the divine creation is praised, through an unceasing thanksgiving and doxology to the all-wise Creator. This doxology has in sight God's manifold gifts to man, but especially the saving sacrifice on the Cross of the God-Man, God's Son and Word, which reveals the incomprehensible efficaciousness of the Cross as the way of transforming and improving the world.

In this way, the natural world acquires deep significance because it participates in the plan of divine economy. It is not a place of exile and imprisonment of the Spirit, but an instrument and garment of it that is being sanctified and is participating in it. The natural world is destined to partake of the renewal and glorification, which encompass the body of the Lord that ascended into heaven. Consequently, preoccupation with nature does not constitute a task that contradicts Christian interests or militates against Christian duties. This presupposes, of course, that such preoccupation is given its rightful place within the context of the rest of the Christian duties, such as, the ministry of the word, or the ministry of the table, the active engagement in good works and every other good work. Having all these things in mind, the Mother Church does not refrain from concern with the problems of the natural environment, knowing that this environment should be of good service to humanity and fulfill the purpose for which it was destined. It is, then, in the context of this interest, that it assumes an initiative and participates in this third international scientific symposium, Religion, Science, and Environment, whose specific theme this year is that of the "Danube—A River of Life."

THE GIFT OF A RIVER

The Danube is a superb gift of God to the person of Central and Eastern Europe because it has been indeed a source of life for all the peoples of Europe. In Roman times, the Danube marked the limit of the civilized world; and in Byzantine times, it was the natural bridge of communication between the peoples of the region and the civilized world. At all times, the Danube has been the open way of constant transportation of material and spiritual goods among the peoples of North and South, of East and West.

Indeed, through the transference of the capital of the Roman Empire from Rome to Constantinople, the great commercial artery of the Danube was connected both with the highest civilization of Byzantium and with the greatest commercial market of the then known world, that of Constantinople. Thus, the Danube served for centuries and to this day, through the richness and natural flow of its waters, both the natural and the spiritual dimension of the life of the peoples of Europe and of the East. This was particularly the case during the Christian period, but it still remains a source and hope of life for the people living beside it and for the people of Europe as a whole.

As a consequence, any indifference toward the vitality of "the river of life" on the part of those near to it or far from it could be described as a blasphemy to God the Creator and as a crime against humanity. This is because the death of its life is a threat to the life of all. The dumping of industrial, chemical, or nuclear waste into the flow of the river of life constitutes an arbitrary, abusive, and certainly destructive interference on the part of humanity in the natural environment. For, through the pollution or contamination of the waters of the river, destruction is procured of the entire ecosystem of the broader region, which receives its life from unceasing communication, like communicating vessels, of the watery subterranean or supraterranean arteries of the earth.

LAWS OF NATURE AND LAWS FOR NATURE

It is obvious, then, that the constantly increasing interest of the European peoples not only for fuller development of the natural element of the river, but also for more direct intervention for the preservation of its natural life, constitutes their supreme duty. This is based on the fact that the life of the Danube is a divine gift for a more effective protection of the life of several European nations. In the opposite case, namely in the case of the continuation of pollution and contamination of the waters of the river of life, the peoples of Europe will destroy a source of their own life for the sake of transient services to insignificant economic or other interests as compared to the divine gift of life.

In light of this, it is clear that the international symposium Religion, Science, and Environment has rightly included in its mission the study of the problem of the Danube and has rightly connected it with sailing through the river of life. The sensitivity of the Orthodox peoples concerning this problem is self-evident; but it has to become a matter of consciousness and personal responsibility for each of us, if it is to be resolved more quickly. In His perfect wisdom, God has laid down the aims and laws that pertain to the operation of the entire divine Creation, and has provided for the self-sufficient protection of its life. Therefore, He designated the human person as a steward, and not as a destroyer of the divine Creation. He did this because humanity is the finest member, the microcosm, the king of the entire divine Creation. Consequently, if humanity's stewardship is unfaithful to the divine commandment, that it should work and maintain the creation within which it was placed, then humanity is unfaithful to itself, destroying God's house, which sustains its own life.

The Ecumenical Patriarchate and the local Orthodox churches—among whom the Most Holy Church of Romania is included, under the inspiring leadership of Your beloved Beatitude, most honorable brother Patriarch Teoktist of Romania[5]—and the entire pious people of Romania, from His Excellency the President of the Republic to the last citizen, are conscious of their mission for the protection of the natural environment. They also know that indifference toward the purpose and the normal operation of the divine creation would be considered today as an unacceptable cosmological stance. This is because the Orthodox Church cannot afford to show lack of concern for the natural world, which was included by God in the plan of the divine economy in Christ. The Orthodox Church knows full well that the renewal and recapitulation of the

---

5. Patriarch Teoktist (1915–2007) became Patriarch of the Romanian Orthodox Church in 1986.

entire creation was envisaged in Christ. Thus, the social realism of the Orthodox faith and the Orthodox dogmatic stance in regard to the creation easily lead to the conclusion that every Christian is both able and obliged to contribute actively not only to the salvation of the river of life, the Danube, but also to the protection of the entire ecosystem of humanity and of the other related ecosystems.

Greeting at Mayoral Reception, Symposium III, Bratislava,
Slovakia, October 20, 1999

TRADE AND NATURE

It is with much delight and great emotion that we find ourselves once again visiting beautiful Bratislava, a city so dear to us, in the context of the international scientific congress on the subject "Religion, Science, and Environment," and this year on the more specific subject "Danube—A River of Life." We recall the beautiful moments that we spent last year during my official visit to Slovakia and this delightful city, and we express our satisfaction because God has given us the privilege of being here again, even for a short while.

We bring to you the blessing and the love of the Mother Church of Constantinople. We address to you our wholehearted greetings in the Lord, and would like to express our affection to you and our gratitude for your love and for all the manifestations of it deriving from your honor toward the Mother Church and to our humble person. We wish all of you every support and assistance from the Lord for all your various efforts and good works.

We are sure that in the list of these good works you have included support for the effort implied by this floating Congress. This effort aims at cleaning up the Danube to remove all pollutants, so that it may be a road of life and not a bearer of death. For, as you well know, the great Danube River is not merely a trade route, used as such for millennia, but also a riverbed over which six to seven thousand tons of water flow each second toward the Black Sea. This water, and that of all the other rivers that flow into it, renews the water of the Black Sea and contributes greatly to preserving life in its ecosystem. In addition, over its long course, the Danube creates many ecosystems of his own, which serve human life in their own way. All these ecosystems are in grave danger of being seriously disrupted, or even completely destroyed because of the thoughtless dumping of wastes and toxic substances into the Danube, thus causing damages not only to the areas right on the river banks, but also to the interior of the countries through which it flows and to even more distant areas.

The object of this present symposium is to study the environmental problems thus created as well as to contemplate practical and affordable solutions to them. It is taking place on the River in order to sensitize public opinion in all the countries involved. Our personal participation aims at emphasizing the moral character of this entire effort. For, in truth, this is a humanitarian effort; and its success depends on the contribution of all our fellow human beings. The only motive that may concern all— and, indeed, we believe that it does—is the moral obligation, the duty to our fellow human beings and to our children and, more generally, to future generations.

As Christians, but also as civilized people, we know and we accept that social coexistence is based on some generally accepted rules, many of which are formulated or amended to address some specific situations, but based on eternal principles. In this case, the eternal principle is "Do unto others as you would have others do unto you," a principle that our Lord expressed in a positive way: "And just as you want others to do unto you, you also should do unto them likewise" (Lk 6.31). And surely, as we do not want other people's wastes and dirt dumped on us, it is our obligation to find ways of not imposing our wastes on others. We are sure that the developed moral sense and the good will of all of you will contribute to finding and applying the best possible solutions so that the Danube remains a river of life and does not become a bearer of death.

Address at Mayoral Assembly, Symposium III, Budapest, October 21, 1999

NATURAL AND HISTORICAL CONNECTIONS

It is with great pleasure and emotion that we have come to the historic and beautiful city of Budapest, capital of the ancient Hungarian people, who next year will be celebrating the one-thousandth anniversary of the foundation of the Hungarian state. We come from the See of the Holy Mother Church, the Great Church of Constantinople, which has very close ties with the Hungarian people. It is a well-known fact that it was through this Church that the Hungarian people became acquainted with Christianity, and that very close ties were developed between the Christian kings of the Byzantine Empire and the Hungarian kings who converted to Christianity, resulting even in blood ties being formed. May we cite the example of the princess Anna, daughter of the Hungarian king Stephen V, who married the Byzantine emperor Andronikos II Paleologos. Moreover, Mary, daughter of the Byzantine king Theodore I Laskaris became the wife of the Hungarian King Bela I. Let us also mention Irene, daughter of the Hungarian king Ladislav I, who married the Byzantine emperor John II Komnenos and was mother of Manuel I Komnenos. Yet even beyond these numerous examples, many Christian Greeks settled in Hungary during the Byzantine period, and their descendants are now Hungarian citizens.

To all of you, then, dear citizens of Hungary, and relatives in spirit and by blood, we bring warm greetings, affection, and the blessing of the suffering Mother Church of Constantinople. We are the bearer of a message of peace, brotherhood, and cooperation among all people for the well being of humankind. Our visit is taking place in the context of the third international scientific symposium on the subject "Religion, Science, and Environment," whose specific theme this year is "Danube—A River of Life." This symposium is taking place on a ship sailing on the Danube, on the two rivers, the Buda and the Pest, on which the beautiful city of Budapest was built. The aim of this symposium is to make the responsible authorities in each country and city, as well as ordinary citizens, aware of the dangers to humanity entailed in the thoughtless pollution of the environment. For, it is not only the major industries that pollute our natural surroundings, but ordinary citizens do so as well. Measures must, therefore, be taken to stop this pollution not only by the state and local authorities but also by every citizen. In addition, the sensitization of the ordinary citizen acts as a source of pressure on rulers and scientists, obliging them to study the situation and take the necessary measures.

NATURAL DAMAGE—HUMAN DANGER

God created the world as something beautiful. Of course, after the original sin of our forebearers, nature became subject to corruption and humanity subject to sin. Nevertheless, through our Lord Jesus Christ,

God renewed His covenant with humanity and nature awaits its liberation from the "bondage of corruption" (Rom. 8.21). However we, and especially we Christians who consider love to be our fundamental duty and an element of our being, have an obligation to make sure that our actions do not become harmful for our fellow men. Pollution and all harmful influences on our environment more generally have an adverse effect on the lives of our fellow human beings, and they must be avoided.

Until a few years ago, the dangers threatening the environment were of no particular interest to the public. Yet, now, these adverse effects have already become evident and dangers to humanity are imminent; for this reason, it is imperative that we all become mobilized. We have placed the program of this particular symposium under the Church's blessing: first, because this is a further sign of our true love for human beings and our desire to help them to improve their standard of living, and second, because we believe that the Church should not be interested solely in the spiritual life of Christians, but also, according to its abilities, in all their needs. This represents both the express commandment of the Lord as well as the long tradition of the Church. And, third, this program has the Church's blessing because interest in our fellow human beings and their various problems is a basic element of higher spirituality, given that it removes people from their shell of individualism and selfishness, and turns them toward their fellow human beings, characteristic of persons with a higher moral and spiritual level.

We are sure that the Hungarian people, who reside along the banks of the Danube, have become fully conscious of the more general importance of keeping the Danube clean both for themselves but also for the peoples living around the Black Sea, into which it flows. Therefore, we are certain that these people will take all necessary means dictated by modern science and technology in order to eliminate its pollution.

Remarks at a Reception by the Greek Ambassador to Norway,
Symposium III, Oslo, June 11, 2002

SCIENCE, ETHICS, AND SACRIFICE

Beyond the technological and scientific dimensions, our symposium widely discussed the ideological perspective of an appropriate environmental ethic. In brief and in summary, we may say that there exist two tendencies. The first of these tendencies demotes humanity and equates it with all other beings in our ecosystem, regarding human survival as equivalent to the survival of any other life form. The second accepts the superiority of humanity over the rest of creation and regards it as a responsible steward looking to preserve creation for the sake of future generations.

Naturally, the second tendency is more correct than the first. This is the tendency that we accept as Christians. However, it is a tendency that conflicts with the harsh reality, according to which, those who are stronger and who pollute the environment are often either deprived of ecological sensitivity, or else hesitate to assume the cost of protective measures for the environment. This is precisely why we are working to sensitize nations, so that the necessity to assume such measures and to discover ways of meeting such costs may become the common conscience of all.

In our concluding address of the symposium, we emphasized that the essential requirement for every good deed is a sense of sacrifice, without which no good may be gained. It is our hope that people will widely appreciate, regarding the natural environment, that which Euripides stated about the human body: "We do not possess this as our own, but we dwell in it during our lifetime."[6] It is our further hope that we care for the environment with the same degree of concern that we do for our children.

Homily at Uspenski Cathedral, Symposium V, Helsinki, Finland,
June 6, 2003

THE IMMORALITY OF INDIFFERENCE

We thank God in the highest, who is worshiped as one in Trinity, for granting to our Modesty the joy also of the present visit to the Holy Orthodox Church of Finland and to its people, irrespective of religious doctrine. We convey to you the warm love and affection, the heartfelt greeting and blessing of the Mother Holy Great Church of Christ of Constantinople. We thank you for the warm reception of our Modesty, for your love and respect for the Mother Church, and for your kind words. The impressions of our gathering today, which takes place on the occasion and in the framework of the fifth international Religion, Science, and the Environment Symposium, will remain indelible in our heart.

---

6. See, for example, *Electra* v. 289.

The institution of these sea-born symposia, recognized by the European Commission and organized under the joint auspices of His Excellency the President of the European Commission and our Modesty, has, as its special theme for this year, the Baltic Sea as a common heritage and shared responsibility of the surrounding countries. The significant number of participants, as well as the scientific and moral authority of these delegates, guarantees the high level of the research, discussions and results.

We participate in these symposia because we believe it is our duty of love toward our fellow human beings to sensitize the conscience of everyone. For, the slightest indifference of each one of us toward our neighbor and the consequences of our actions has led to the current critical situation of the excess pollution of the seas, especially the closed ones, as well as of many regions of the land. This indifference has disturbed the ecosystems and in many cases caused their destruction, with very unfavorable results for humankind, such as the destruction of fisheries and agriculture in some places, the pollution of the atmosphere and waters by toxic materials, and other reversals of the natural environmental balance, which is necessary for the regular and healthy life of humanity. From the Christian perspective, it is unethical to be indifferent toward the increasingly negative repercussions of our actions simply because each of them has a very small impact on the situation. Experience has shown that, beyond a certain degree of self-purification, the environment cannot recover its natural condition and is gradually dying.

We are obliged to become conscious of the truth concealed in the teaching of the Apostle Paul that we all constitute one body, and when one member suffers, all members suffer together. Any form of individualism, which leads us to care only for ourselves, is antichristian. We who believe in Christ have to live the truth that we are all one body, that we all have a common interest, and that we must seek not only our own, but also the interests of every person (Phil. 2.4).

Remarks in the State House, Symposium VI, Manaus, Brazil,
July 19, 2006

THE AMAZON IN CRISIS

It is a privilege and a joy to be received as your guest in this historic city, the capital of an extraordinarily beautiful, and infinitely precious part of the earth known as the state of Amazonas. As administrator of a large and exceptionally rich section of God's creation, you shoulder a responsibility, which would be hard for any politician in the so-called developed world to imagine. The natural environment under your care is the finest and most perfectly intact part of the Brazilian rainforest, a biological and ecological treasure-house, on whose survival the world depends. The people under your care represent, in an extreme form, the radiance, diversity, and passion of the Brazilian nation as a whole. They range from indigenous communities who have had little or no contact with the supposedly civilized world, to poor and vulnerable city-dwellers who have recently arrived in this region in the hope of making a living. For the simple fact that that you and your state ministers have undertaken the governance of such a rich and yet fragile part of our planet, we salute you and we respect you. We especially admire your state's commitment to preserving as much as possible of the rainforest, and to protecting the indigenous people who are guardians and guarantors of the forest's survival.

The administration of such a place would be a difficult task in normal times, but these are not normal times. Last year, the world learned with distress and amazement of the crisis with which your state had to cope. In a region that we would normally associate with abundant supplies of water, there was a drought that killed millions of fish and left many communities with nothing safe to drink. In a part of the world where nature offers such a huge range of edible produce, people were left short of food. In remote areas where people rely on river transport to sell their own crops and procure whatever food and medicine they need, there was great distress because local waterways dried up. We recognize the resourcefulness of the Amazonas state authorities, and of all the Brazilian authorities, in dealing with this crisis, which many scientists saw as a grim warning of a broader breakdown in nature's equilibrium.

CAESAR AND GOD

From the Ecumenical Patriarchate, endowed by history and tradition with a moral obligation toward the whole of creation, we offer our love, fatherly concern, and fervent prayers for all the people involved in administering the state of Amazonas, as they cope with the human consequences of what may turn out to be a continuing environmental crisis. If we hold back from offering any more detailed words of counsel, that is in part because we are guided by the arresting command of our Lord that we should "render unto Caesar the things which are Caesar's and unto God the things that are God's" (Mt. 22.21).

Many commentaries on those famous words have pondered the precise meaning of "rendering unto Caesar" our obligations as citizens. However, it is also worth reflecting seriously on what is meant by "rendering unto God" what is due to God. At first sight, this seems to be a reminder that we should carry out our formal religious obligations, by praying, fasting, going to church and observing the church calendar. That may certainly be one part of what our Lord is telling us. But in the course of the five previous symposia, which we have organized as part of the movement created by Religion, Science, and the Environment, we have tried to look more deeply into what it means to offer unto God what belongs to God. We have reflected on man's role as a priest of creation, a creature with a unique calling to receive all the bounty of the created world as a gift from God, and then to offer that gift back to God in a spirit of gratitude and humility. In some form or another, almost every human culture has felt the impulse to offer the treasures of creation, including life itself, back to the Creator. As Christians, we have a particular sense that in making our eucharistic offerings of bread and wine, we are joining or becoming part of the supreme, once-and-for-all sacrifice made on the Cross by the Son of God as an ultimate act of saving love.

But whatever we believe about the nature of priesthood or sacrifice, it is worth noting that modern, secular man is out of step with most of human history in one important way: in his view that the created world is merely something to be exploited and abused with no sense of respect for a Creator or even for future generations. This absence of gratitude or respect is as shocking to our Orthodox Christian tradition as it is to any of this region's peoples, who share with us the unshakeable belief that all forms of life are deeply connected, and that every living thing has its own divinely ordained purpose, its own logos as we say in Greek. Among people who are guided by those principles, certain other things should follow: the need to show respect, prudence, and self-restraint in our treatment of the created world, in short to love the created world, just as we are united in love to our Creator. Unless that understanding exists and is deeply felt, government policies alone, however prudent and wise, will hardly be able to save the planet from destruction.

Respected governor, we commend and admire the courage and competence you have shown as a person who is called to play the role of Caesar, in other words to look after a large and exceptionally challenging piece of territory for the benefit of its people. For our part, we at the Ecumenical Patriarchate will strive, as far our human strength allows us, to bear witness to another important truth, namely that no earthly administrators can succeed in their mission unless people are also willing to offer back to God all those things—especially the great and wondrous gifts of nature—which came from God and ultimately belong to Him.

5

Prayer and Spirituality

Transfiguration and Sacrifice

Opening Ceremony, Ecological Symposium, Halki, Turkey, June 1, 1992

JOINT PRAYER AND COLLECTIVE ACTION

As far as our most holy Orthodox Church is concerned, we have reasons to be very intensely concerned for the protection of the natural environment. This was demonstrated during the Inter-Orthodox Conference convened, at the initiative of our Throne, last November on the island of Crete, which Your Royal Highness[1] also honored with your personal presence and active participation. These reasons, then, for our concern may be distinguished into two basic categories: There are theological reasons, reasons of faith; and there are also pastoral reasons, reasons of sensitivity toward the world, and to the mission and service of the Church in this world.

As to the first category, namely the theological reasons, it is known that the Fathers of the Church always perceived salvation in Christ as relating not only to humanity, but through the human person also extending to all of creation. Our Lord Jesus Christ is the "recapitulation" of all of creation, according to the well-known saying of St. Irenaeus, Bishop of Lyons in the second century.[2] St. Maximus the Confessor, the

---

1. Prince Philip, Duke of Edinburgh.

2. Irenaues (d. 202) was one of the earliest theologians and apologists of Christianity, with direct spiritual lineage to the apostolic community.

great seventh century theologian and Father of the Orthodox Church, was so insistent upon the importance of material creation in the whole divine plan of salvation that he saw the human person as a microcosm, and considered salvation in Christ as a "cosmic liturgy" within which all of material creation participates. Moreover, it is indicative of the faith of the Orthodox Church that, at the very epicenter of her life lies the divine Eucharist. This Eucharist is nothing else than a "liturgy," namely a communal act of the people of God, wherein the faithful offer the gifts of creation, the bread and the wine, in order that these may be changed into the Body and Blood of Christ, and be returned in thanksgiving to the Creator. Orthodox Christians firmly believe that God not only created the material world, but that, through humans, the created world also has, as its destiny, the participation in the glory and eternal life of Christ.

Consequently, the final purpose of creation is not its use or abuse for man's individual pleasure, but something far more sublime and sacred. It is from these points, then, that the pastoral reasons for the Church's concern for the natural environment also emanate. For us Orthodox, every destruction of the natural environment caused by humanity constitutes an offense against the Creator Himself and arouses a sense of sorrow. In relation to the degree to which people are responsible for their actions, *metanoia*—a radical change or course is demanded of us all. For this reason, each human act that contributes to the destruction of the natural environment must be regarded as a very serious sin. We are talking here about a renewed ethos, which must be taught to our faithful. Our faithful must become sensitized to the gravity of this sin and to the need to espouse a corresponding ethos. People must cease regarding themselves as *proprietors of nature* and understand their mission as *priests of creation* who have as their duty the *anaphora*, or offering up of the material world to the Creator. In this new ethos, the liturgical and the ascetic tradition of the Church can be of assistance to its faithful.

Remarks at a Banquet in Honor of HRH Prince Philip,
Duke of Edinburgh, Istanbul, Turkey, May 31, 1992

A COMMON OBLIGATION

It is with great joy and profound esteem that the Ecumenical Patriarchate, in our humble person and on behalf of the Holy Synod, welcomes Your

Royal Highness here today to this sacred Center of Orthodoxy. Your visit brings special honor to our Church here as well as to the entire Orthodox Church. We thank you for the effort that you so kindly incurred in order to come here, and for willingly responding to the invitation of our predecessor, Patriarch Dimitrios of blessed memory, which our Modesty very happily renewed.

The bonds between Your Royal Highness and the Orthodox Church are old and deep. Deep and close, as well, are the bonds that for a long time have bound the Orthodox Church, and particularly the Ecumenical Patriarchate, with the Church of England, whose head is Her Majesty Queen Elizabeth II. Your presence here further strengthens these bonds, promoting unity among peoples and among human beings, a unity of which the contemporary world is in much need.

A UNIQUE WITNESS

We are particularly pleased by the fact that Your Royal Highness is visiting this sacred center after so short a time since the assembly of the Inter-Orthodox Committee was convened on the Island of Crete at the Orthodox Academy in order to study the issue of the protection of the natural environment. All of the Orthodox Churches, particularly the Ecumenical Patriarchate, were moved by the fact that Your Royal Highness deemed fit to honor this conference with your participation and to inaugurate its sessions with your inspired address, which was appreciated most deeply by the participants at the conference as it indeed added inestimable meaning to its success. Your visit here offers us the opportunity to express our gratitude for Your Royal Highness's contribution toward so delicate and sensitive an initiative as that taken by the Ecumenical Patriarchate.

The sensitivity manifested by Your Royal Highness toward the protection of the natural environment is well known and deeply appreciated by all. As President of the World Wide Fund for Nature International (WWF), you have made an inestimable contribution toward the preservation of nature. This contribution on your part is especially moving to our Church, which considers the preservation of God's creation to be a part of her sacred duty. As you have been made aware from the encyclical message of our predecessor Patriarch Dimitrios of blessed memory (dated

September 1, 1989),[3] our Church has designated the first day of September of each year as a day of prayers for the natural environment. It is our hope and prayer that this proposal by the Ecumenical Patriarchate, which has already been accepted by the Primates of all the other Orthodox Churches during their recent assembly, which took place here on the Sunday of Orthodoxy, might be adopted, if possible, by all Christians in general. This would contribute greatly to the establishment, by all believers in Christ, of a common and unique position on so critical an issue as the protection of God's material creation.

Opening Address at a Conference on the Natural Environment,
Mt. Athos, Greece, September 29, 1997

THE PATRIARCHATE AND THE ENVIRONMENT

Announcing the commencement of the proceedings of a two-day conference on the natural environment of the Holy Mountain, we take the opportunity to expound certain basic theses related to our interest in the natural environment.

The sensitivity of the Ecumenical Patriarchate for the environment reveals the sensitivity of the Great Church of Christ for human beings. This is because the environment is not viewed per se, but in relation to them for whom it was made. Hence, every discussion about the environment and every effort of preserving or improving it indicate interest in the conditions of human life. Even in cases where this is not readily understood, or where human interests in self-survival seem to be conflicting with human interests in maintaining the environment, the ultimate interest of man, or, perhaps, tomorrow's man, demands that priority is given to the latter over the former. But we should never forget that the aim in all this is to improve the conditions of human life because man is the king of creation, and creation was made to serve man's needs. We do not idolize the natural environment, but we try to accentuate man's sensitivity to the conditions of life of his fellow man.

Drawing our inspiration from such a compassionate spirit, we salute the rise of general interest and of discussions on the natural environment as expressing curtailment of individualism and expansion of altruism,

---

3. See Chapter 1, "Call to Vigilance and Prayer."

which constitute a basic precept of the Gospel of Christ. Hence, it is with joy and deep sentiment that we announce the commencement of the proceedings of this two-day conference for the natural environment of the Holy Mountain, which was summoned by the Organization of the Cultural Capital of Europe (Thessalonika, 1997), in the context of the activities of the exhibition *The Treasures of the Holy Mountain.*

THE SILENCE AND BEAUTY OF MT. ATHOS

This context, within which the present two-day conference takes place, prompts our Modesty to draw everyone's attention and especially of the most sacred monks of the Holy Mountain and of every friend of Athos, as well as of every responsible person, on the need to protect the special importance of the natural environment of the Holy Mountain.

It is well known that the natural environment does not belong only to the present generation but to the future generations as well. Indeed, the present generation has received it as a torch and is obliged to pass it on to its successor, if there is in its heart the love for it. This, of course, is also applicable to the natural environment of the Holy Mountain.

There are not a few regions of the earth, where, for the sake of a more general interest, which stands in opposition to the narrowly defined interests of the local inhabitants of these regions, proscriptions of environmental interventions have been imposed. Thus, for example, carving roads and introducing vehicles have been proscribed for the Island of Hydra in Greece and the Princes' Islands in Turkey. And as the *Sayings of the Desert Fathers* report, a certain ascetic built his cell far away from the spring in order to exercise himself laboring for the transportation of the necessary water. This means that it would not be excessive if one were to demand of those who chose the monastic life out of their own volition to care less for convenience and more for the preservation of the natural beauty and the silent character of the Holy Mountain. Indeed it is this quietist character that attracts those who are pushed away from the whirl of social life to enter monasticism, and the possibility of finding the same whirl at this place of quietism may cause problems for them in their new pursuit. There is, then, a man-loving duty and a cultural need to preserve unaltered the natural environment of the Holy Mountain, at the cost of some small sacrifices of material conveniences on the part of its present inhabitants for the incomparably greater spiritual convenience of those who are to come there in the future. Indeed, there is no one that ascended to heaven with ease, as Abba Isaac says who is much admired by the most sacred monks of the Holy Mountain. And in order to sharpen somewhat our discourse, we remind you with paternal love of the exact saying of Abba Isaac: "God and His angels rejoice where there are needs, but the devil and his friends do so at times of ease." And again in order to alleviate the sharpness of the reminder, we affectionately quote from the same Abba a message of consolation: "To any discomfort undergone for God there is always a comfort that follows."

We may also conclude, then, that to any discomfort that we may undergo for the sake of our neighbor, discomfort caused to us and to those around us by caring for the environment, there will always follow a comfort as a counterpart. This is exactly what we wish for all of you and furthermore we wish a good yield from the proceedings of this two-day conference.

Keynote Address, North Sea Symposium, Utstein Monastery, Germany, June 23, 2003

THE ASCETIC CORRECTIVE

This session marks the opening of the Sailing Seminar on the North Sea. It takes place within the walls of a strategically placed monastery at the entrance of the magnificent and unique Ryfylke fjords, where the marine traffic along this coast was once controlled. Inhabited as an island from as early as the Bronze Age, it has been a special haven for monks of the Augustinian Order since the Middle Ages. This monastic setting surely provides for us an ideal opportunity to assess the importance and impact of the phenomenon and experience of monasticism in general for the ecological balance of our world.

At the conclusion of the service for the tonsure of an Orthodox monk or nun, the newly received member of the monastic brotherhood or convent stands before the entire community bearing three simple tokens: a cross, a candle, and a prayer rope. The first two symbols—the cross and the candle—standing as we are today in this historic and royal monastery of Utstein, are a powerful reminder of the ecological corrective offered by the monastic way of life. The monk and the nun, representative of all the Christian faithful, and indeed of the whole world, are an image of the spirit of asceticism or self-restraint and of a light that illumines and protects the world in the face of every form of spiritual darkness.

THE POWER OF THE CROSS

In his letter to the Colossians, St. Paul writes:

> Through him [Christ], God was pleased to reconcile to himself all things, whether on earth or in heaven, by making peace through the blood of his cross. (1.20)

Reference here to "the blood of the cross" is a clear indication of the cost involved in any efforts that we might undertake to address environmental problems of our time. The cross is the singular, ultimate, and absolute solution to the ecological crisis. The cross reminds us of the reality of human failure and of the need for a cosmic repentance. In order to alter our attitudes and lifestyles, what is required is nothing less than a radical reversal of our perspectives and practices. There is a price to pay for our wasting. It is the cost of self-discipline.

This is the sacrifice of bearing the cross. The environmental crisis will not be solved simply by sentimental expressions of regret or aesthetic formulations of a creative imagination. It will not be altered by fashionable programs or ecumenical catch words. It is the "tree of the cross" that reveals to us the way out of our ecological impasse by proposing the solution of self-denial, the denial of selfishness or self-centeredness. It is, therefore, the spirit of asceticism that in the final analysis leads to the spirit of gratitude and love, to a rediscovery of the sense of wonder and beauty.

THE WAY OF ASCETICS

In this context, we would define asceticism as the possibility of traveling lightly, of using and consuming less. And we can always manage with much less than we imagine. We are to learn to relinquish our desire to possess and control. We must stop wounding the natural resources of this earth and learn to live simply, no longer competing against one another and against nature for our survival. What is called for is a softening up in our relations toward each other and toward nature. We must learn to make our communities more sensitive and to render our behavior toward nature more respectful. This means acquiring a merciful attitude, a compassionate heart. Such a heart cannot bear to deplete—still less to destroy—the earth that we inhabit and share. In the seventh century, St. Isaac of Syria defined this as:

> Having a heart that burns with love for the whole of creation: for humans, for birds, for beasts, even for demons—for all God's creatures.[4]

Asceticism, then, aims at a sense of refinement, not at any form of detachment or destruction. Its goal is always moderation, never repression. The content of asceticism is positive, not negative. It looks to service and not selfishness, to reconciliation and not renunciation or escape. Without asceticism, none of us is authentically human. Without asceticism, none of us can hope to heal our broken environment.

The general impression that people in western societies have of asceticism is negative. Asceticism carries with it the baggage of dualism and denial, developed over many centuries, both inside and outside the Christian church. This is why so many people have misunderstood and even dismissed monasticism. Yet this is not the vision of wholeness that Orthodox spirituality proposes through its ascetic dimension. The sacramental dimension of the world is intimately and profoundly linked with the ascetic dimension. Asceticism is the conscious awareness and deeper recognition that humanity is dependent not only on God, but also on the world, and indeed on the food chain, just like every other creature made by God.

Such asceticism, however, requires from us a voluntary restraint in order for us to live in harmony with our environment. Asceticism offers practical examples of conservation. By reducing our consumption—what in Orthodox theology we call *enkrateia* or self-control—we come to ensure that sufficient resources are also left for others in the world to share and enjoy. As we shift our will and focus of concern, we shall be able to demonstrate a compassion for the poorer nations of our world. Our

---

4. *Ascetic Treatises* 48.

abundance of resources should also be extended—beyond ourselves and our own—to include an abundance of equitable concern for others.

This further implies that humanity is not to act as the tyrannical over-lord but as a servant and minister, who kneels in prayer for the preservation and progress of creation. In this way, humanity is able to restore harmony with the rest of the world, as well as to reconcile all people and all things with God. This responsibility or obligation underlines the priestly dimension of the human vocation. Human beings are called to be priests and not proprietors of nature. Humanity has an active role to play within the world, endowed with the moral responsibility to assume creation in an act of giving in order to refer it to God in an act of thanksgiving.

The seventh-century hermit on Mt. Sinai and author of *The Ladder of Divine Ascent*, St. John Climacus, who is remembered every year in Orthodox monasteries during Great Lent, wrote: "A monk without possessions is master of the entire world." Similarly, St. Paul recommends the avoidance of avarice, when he writes: "As we have food and clothing, let them suffice to us" (1 Tim. 6.8). Asceticism and self-restraint are ways of realizing the words of St. Paul, who elsewhere says that we are to be: "As having nothing yet possessing all things" (2 Cor. 6.10). It is following the commandment of Christ:

> Whoever wishes to save his life shall lose it; but whoever loses it for my sake shall find it. (Matt. 16.25)

Now, this voluntary ascetical life is not required only of the hermits or monastics. It is also demanded of all Orthodox Christians, according to the measure of balance. That is to say, each Orthodox Christian is called to practice a voluntary self-limitation in the consumption of food and natural resources. Each of us is called to make the crucial distinction between what we want and what we need. Only through such self-denial, through our willingness sometimes to forgo and to say "no" or "enough" will we rediscover our true human place in the universe. Such is part and parcel of the ascetic ethos of Orthodox spirituality.

FASTING AS CORRECTIVE TO WASTING

Let us, by way of example, explore one specific aspect of ascetic practice in the Orthodox Church, namely fasting. Orthodox faithful fast from all dairy and meat products for almost half of the entire year. There is the great fast of Lent before Easter, the Lenten season prior to Christmas, a period of fasting before the feasts of the Dormition in August and of the Apostles in June. There is also a weekly discipline of abstaining from meat and dairy produce each Wednesday and Friday of the year (monastics additionally fast every Monday), as well as certain specified days of fasting in commemoration of particular events or saints. The fact that we fast almost half of the year is itself an indication of an effort to reconcile secular time with sacramental time, the time of this world with the age of the kingdom to come.

If we examine fasting more carefully, we shall see that it does not aim at denying the world, but at affirming the significance and sacredness of material creation. It is a way of integrating body and soul, while at the same time recalling the hunger of others. Indeed, it is becoming aware of the hunger of creation itself for healing and restoration. When we fast, we remember that we live not by bread alone. We also remind ourselves that we are a part of an entire community, the fellowship of humanity, and the natural environment. This is why Orthodox Christians will never fast alone or at whim; we always fast together and at set periods of the year.

Therefore, fasting is the conviction and acknowledgment that "the earth, is the Lord's, and all the fullness thereof" (Ps. 23.1). It is an affirmation that the material creation is not under our control; it is not to be exploited selfishly, but rather to be returned in thanks and restored in an act of communion with God. In the final analysis, and beyond people's perception of fasting and asceticism, fasting is learning to give and not simply to give up. It is learning to share and to connect with others and with the natural world. It is a way of loving, of moving gradually away from what we want to what God's world needs. It is liberation from fear, greed, and compulsion. It is regaining a sense of wonder, being filled with a sense of goodness, and seeing all things in God, and God in all things.

The Orthodox attitude of asceticism and practice of fasting appears not to impose any method of dealing with and solving the environmental problems of our time. However, just as the individual actions of tens of thousands of members of society produce great pollution, so the voluntary restraint of an Orthodox monk is of great benefit for all. It is a lesson in "cultivating and keeping the earth" (Gen. 2.15). Therefore, the discipline of fasting becomes a necessary corrective for our culture of wasting. It is a way of breaking bad habits and of seeing the world with new eyes, or with God's eyes.

Now, by way of conclusion, you will perhaps remember that, at the outset of my address, we mentioned that the newly tonsured Orthodox monk or nun stands before the community bearing three symbolic tokens. The last of these is a prayer rope. It is a symbol of the continuous struggle and desire expressed through prayer for the protection of our world. The environmental program of the Churches cannot simply involve philosophical or political changes. It must include a spiritual repentance that occurs only through unceasing prayer. Addressing the ecological issues of our time will not only take place in the public sphere or the political domain. It will primarily take place on our knees.

Remarks During the North Sea Symposium, Institute of Marine Research, Bergen, Norway, June 24, 2003

THE SIGN OF JONAH

When the scribes and Pharisees once asked our Lord for a heavenly sign, He warned them by way of response, stating:

> This generation asks for a sign, but no sign will be given to it except the sign of the prophet Jonah. (Matt. 12.38–39)

It would be wise for us, therefore, in our genuine search for a sign in regard to the role that the churches have to play in the environmental crisis that we face, to heed these words of Christ. They contain the seed of our contribution toward a solution, without which we would only continue to be a part of the problem. Christ refers to a sign, to the power of prophecy, and to the person of Jonah.

A REMARKABLE ICON

There is a profound iconographic depiction of this seed in an eighteenth-century icon at the Monastery of Toplou in Crete.[5] (See page 206.) The

---

5. The icon may be found at the Holy Stavropegic Monastery of Panagia Akrotiriani in Toplou, Seteia, on the island of Crete. Its dimensions are 55″ × 41″. For more images,

iconographer of this sacred image is Ioannis Kornaros.[6] It is truly a theological statement in color. The icon assumes its title from a prayer found in the Great Blessing of the Waters chanted during the Feast of Epiphany on January 6 each year, and also repeated during the baptism of every Orthodox Christian:

> Great are You, O Lord, and wondrous are Your works; no words suffice as
> a hymn to praise Your wonders!

At the far left of this image, created nature is portrayed as a woman, reflecting the concept of "mother earth" that indigenous peoples throughout the world—be it the Indians of North America or the Aborigines of Australia—have respected and retained for centuries. Nature holds her arms open in a gesture of receptiveness and embrace of all people and all things.

The icon also reveals the reality of urban life (with the scene of two cities, that of Samaria and that of Nineveh, in the background) as well as of agricultural life (with farmers tilling the soil on the mountain slopes). There are rivers and there is vegetation. Human beings are shown beside a number of animals of a wide diversity, while a vast rainbow extending over much of the icon itself reflects the eternal covenant between the Creator and His creation.

THE SACREDNESS OF FISH

While this icon is abundantly rich in symbolism, sacredness, and significance, there are two scenes that are of particular value for our discussion of fisheries in the North Sea. The first of these scenes depicts Jonah being cast out of the mouth of a large beast of the sea, in accordance with the biblical story. This is of course a powerful and profound image of the resurrection of life and the renewal of all things by the Risen Lord, who

---

see T. Provatakis, *Icons of the Painter Ioannis Kornaros at the Toplou Monastery of Crete*, Toplou: Holy Monastery of Akrotiriani, 1984.

6. Though little is known of the life of this Cretan iconographer, Kornaros lived from 1745–1796 and painted this icon in 1770.

The icon entitled "Great are You, O Lord" at Toplou Monastery in Crete

desires the salvation of all people and the life of the whole world. As it is well known, one of the early symbolisms of Christ, through which Christians recognized and greeted one another, was the sign of the fish. In Greek, fish is denoted by the word *ichthys*, a term whose initials spell out the phrase: "Jesus Christ, Son of God, Savior."

The fish, then, is a soteriological statement of faith. Christ has been intimately and integrally identified with the fish of the sea. Therefore, any misuse or abuse of fishing and fisheries relates in a personal and profound way to Christ Himself. It leaves a scar on the very Body of Christ Himself. The image of the fish has from the earliest of Christian times also been appreciated and accepted as a prophetic sign of salvation and resurrection. The Church, then, is called to protect the fisheries of the world and to speak with clear voice and prophetic criticism against any kind of over-fishing and every form of pollution of the world's fisheries. The way we treat our fish in the waters ultimately reflects the way that we worship our God in the heavens. Fishery and liturgy are closely connected and inseparably interrelated.

CAIN AND ABEL

The second scene depicts the slaying of Abel by Cain.[7] This is a violent and cruel representation of the negative and destructive impact that our current practices and policies bear upon future generations. We can no longer remain passive observers—or, still worse, active contributors—to the merciless violation and destruction of the natural environment and its resources, as well as to the unnecessary and unbearable extinction of the numerous and diverse species of flora and fauna.

Until we can perceive in the over-fishing of our seas the very portrait of our brother and sister, we cannot hope to resolve the inequalities of our world. Indeed, until we can discern in the pollution of our fisheries the very face of our own children, we shall not be able to comprehend the immense and lasting consequences of our attitudes and actions. We must place limits on our insatiable desires that are so obviously encouraged by the prevailing philosophy of the consumer society that we have learned to take for granted.

The consumption of the resources of our seas is more than a matter of dietary or culinary delicacy. It is a matter of social justice and delicate behavior toward our neighbor. The riches of the North Sea—its fisheries and sea farming—are not a matter of commercial or financial development. They are a spiritual problem that affects the very survival of our

---

7. See Genesis 4.

children and of our planet. We cannot even imagine a future for our planet without some form of sustainable—or, more correctly, restrained—development.

ICONS AND CREATION

Finally, the very notion of icons introduces yet another aspect of our relationship to the material creation and the natural environment. We all know that this world was created "good," indeed "very good," by a loving Creator, who has permitted and even commanded us "to serve and to preserve" (Gen. 2.15) the resources of our planet. Yet, we are called to do more than simply preserve nature or conserve the world. It is inappropriate for us either to control or manage God's creation in a utilitarian way or even to conserve the natural environment in a passive way. Nature is not an object for our use or abuse.

Human beings are endowed with the distinctive gift of freedom. It is the freedom to create a new environment. It is the possibility to assume this ephemeral world and to render it eternal by referring it back to its source and creator, namely to God. Human beings are, therefore, called to transform nature into culture. This is precisely what happens in the event of the icon, which takes the material resources of wood and egg and paint, finally transfiguring these through prayer into a sacramental encounter and experience of the living God. In this way, humanity embraces nature and does not simply manage it. In this way, the whole world assumes the proportions of a cosmic liturgy, where each of us individually—and at the same time, all of us communally—celebrates the gift of life that sustains the whole world.

This communal or liturgical dimension underlines the critical ecumenical imperative of our contemporary ecological endeavors. If we are—as individuals or as churches—to behold the sign of Jonah, humility is a necessary prerequisite and condition. Ministers, politicians, scientists, media, and every member of public and civil life are required to look to and work with each other for answers. Together, we can hope and pray for the healing of the world. Alone, we can only wound and worsen the situation. No individual person, no single society, and no isolated church or religion can hope to resolve or reconcile the crisis that we are facing without cooperating with all others, with all other professions, and with all other disciplines.

Homily at Vespers, Bergen Cathedral, Norway, North Sea
Symposium, June 24, 2003

VESPERS AND GENESIS

During every Orthodox service of Vespers, the opening Psalm (103 LXX) is a song chanted to God as Creator and sustainer of all creation. The service normally takes place at the setting of the sun, when the light of this world begins to wane and the light of God must be yearned. The impending darkness of the night is symbolic of the original moment of Genesis, when "the earth was a formless void and darkness covered the face of the deep" (Gen. 1.2). It is then, as we are told in the first book of Scripture, that "the spirit of God swept over the face of the waters; and God said, 'Let there be . . .'" (Gen. 1.2–3).

Thus, during the vesperal service, we recall the primal darkness that received the first light of God's grace and presence. And we bless God with all our soul and in joyful song, saying:

> Bless the Lord, O my soul. O Lord my God, you are very great. You are clothed with honor and majesty, wrapped in light as with a garment. You stretch out the heavens like a tent, and set the beams of your chambers on the waters. . . . You make the winds your messengers, fire and flame your ministers. (Ps. 103.1–4)

Ultimately, we are recognizing God as the Creator of all things. Indeed, during this seminar, we also recognize the striking beauty of the North Sea in the words of the same Psalm:

> You set the earth on its foundations, so that it shall never be shaken. You cover it with the deep as with a garment; the waters stand above the mountains. . . . You make springs gush forth in the valleys and flow between the hills. . . . From your lofty abode, you water the mountains; the earth is satisfied with the fruit of your work. . . . O Lord, how manifold are your works. In wisdom you have made them all. (Ps. 103.5–24)

Then, the Psalmist offers us a subtle but significant hint of the mutual love and reciprocal exchange that occurs between God and His creation. This he describes in a movement of looking:

> The Lord delights in His works; He looks upon the earth. (Ps. 103.31–32)

God cares deeply about this world. He is concerned about the welfare and future of His creation. He never abandons the work of His hands—both the natural environment and the human person made in the divine image and likeness (Gen. 1.26). This merciful love is revealed in the way that God never ceases "to look upon" this world. His gaze forever accompanies us.

"ALL THINGS LOOK TO GOD"

By the same token, humanity must neither abuse nor abandon the created world. The human person is called—in his or her capacity as priest of creation and never as proprietor of its material resources—to refer to God as the source and center of all beings and all things. Thus, as the Psalmist observes so eloquently:

> All things look to you, Lord . . . you renew the face of the earth. (Ps. 103.27–30)

Again, the gesture is one of "looking to" God. God's eyes meet our eyes and the eyes of the world in an act of communion and love.

Finally, this mutual exchange, the giving and the receiving, the one who loves and the one who is loved, coincides in the unique person of Jesus Christ, who is fully divine and fully human. When we remember our calling to return all things in thanksgiving to God, who renews all things for us in love, then—as the Orthodox service of Vespers concludes in the words of an ancient hymn from the early, undivided Church—we are able to behold:

> The radiant light of the sacred glory of Jesus Christ . . . singing praises at all times [and in all places] with joyful voices, just as the entire world offers glory to God.

Opening Address for the World Exposition, Aichi, Japan,
September 20, 2005

MEDIATING FOR SUSTAINABLE DEVELOPMENT

There is a profound relationship between the divine Creator and the natural creation—worshipping the former and venerating the latter. The future of this planet is of critical importance for the kingdom of heaven. Our understanding of salvation is not other-worldly but involves the transfiguration of this world in light of the heavenly kingdom. This is why, over the last two decades, the Orthodox Church has prayed throughout the world for the protection and preservation of the natural environment. It is critical that the intimate connection between poverty and the natural environment be recognized if problems of either economy or ecology are to be addressed. The natural environment cannot be separated from personal piety and spirituality.

Climate change and environmental pollution affect everyone. While the data may be variously debated, the situation is clearly unsettling. To take but one example: Dramatic increases of greenhouse gases in our atmosphere—largely due to fossil fuel burning—are causing global warming and in turn leading to melting ice caps, rising sea levels, the spread of disease, drought, and famine. The European heat-wave of 2003 could be unusually cool by 2060, while the 150,000 people that the World Health Organization conservatively estimates are already dying annually because of climate change will be but a fraction of the actual number.

It is painfully evident that our response to the scientific testimony has been generally reluctant and gravely inadequate. Unless we take radical and immediate measures to reduce emissions stemming from unsustainable—in fact unjustifiable, if not simply unjust—excesses in the demands of our lifestyle, the impact will be both alarming and imminent.

Religious leaders throughout the world recognize that climate change is much more than an issue of environmental preservation. Insofar as it is human-induced, it is a profoundly moral and spiritual problem. To persist in the current path of ecological destruction is not only folly. It is no less than suicidal, jeopardizing the diversity of the very earth that we inhabit, enjoy, and share. Indeed, we have repeatedly described it as a sin against both the Creator and creation. After all, a handful of affluent nations account for two-thirds of global GDP and half of all global carbon dioxide emissions.

ENVIRONMENT AND SPIRITUALITY

Ecological degradation also constitutes a matter of social and economic justice, for those who will most directly and severely be affected by climate change will be the poorer and more vulnerable nations (what Christian Scriptures refer to as our "neighbor") as well as the younger and future generations (the world of our children, and of our children's children). Those of us living in more affluent nations either consume or else corrupt far too much of the earth's resources.

There is a close link between the economy of the poor and the ecology of the planet. Conservation and compassion are intimately connected. The web of life is a sacred gift of God—ever so precious and ever so delicate. We must serve our neighbor and preserve our world with both humility and generosity, in a perspective of frugality and solidarity alike. After the great flood, God pledged never again to destroy the world: "As long as the earth endures, seedtime and harvest, cold and heat, summer and winter, day and night, shall not cease" (Gen. 8.22). How tragic, however, it would be if we were the ones responsible for their destruction. The footprint that we leave on our world must be lighter, much lighter.

Africa is the continent least responsible for global warming, yet bearing the most detrimental consequences, while also being the least equipped to cope with the changes. Harvest cycles in Ethiopia and other parts of eastern and southern Africa are shortening, leading to further food insecurity for the world's poorest people. Elevated temperatures create incalculable increase in the range of vector-borne diseases and lack of clean water. Populations affected by fatal diseases, such as malaria, schistosomiasis, dengue fever, and cholera, are rising dramatically. Even a conservative estimate indicates the number of people impacted by flooding could increase from 1 million (in 1990) to 70 million (by as early as 2080).

Faith communities must undoubtedly first put their own houses in order; their adherents must embrace the urgency of the issue. This process has already begun throughout the world, although it must be expanded and intensified. Religions realize the primacy of the need for a change deep within people's hearts. They are also emphasizing the connection between spiritual commitment and moral ecological practice. Faith communities are well-placed to take a long-term view of the world as God's creation. In theological jargon, that is called "eschatology." Moreover, we have been taught that we are judged on the choices we make. Our virtue can never be assessed in isolation from others, but is always measured in solidarity with the most vulnerable. Yet churches, mosques, synagogues, temples, and other houses of worship consume a fraction of energy compared to manufacturing industries, modern technologies, and commercial companies.

Breaking the vicious circle of economic stagnation and ecological degradation is a choice with which we are uniquely endowed at this crucial moment in the history of our planet. The destructive consequences of indifference and inaction are ever more apparent today. At the same time, the constructive solutions to mitigate global warming are increasingly merging. Therefore, government, businesses, and religious institutions are actively pursuing cooperation, converging in commitment and compelling people to act. Only together—in a genuine spirit of dialogue and cooperation—can a problem of this magnitude be addressed and resolved. The responsibility as well as the response is collective. We know how to prevent such rapid climate change. We now need a powerful network of diverse yet connected leaders committed to empowering individual and commercial action.

THE PRACTICE OF FASTING

In the Orthodox tradition, the connection between ecology and spirituality is especially tangible in the practice of asceticism, and indeed most notably in the act of fasting. The ascetic way is primarily a way of liberation. And the ascetic is the person who is free, uncontrolled by attitudes that abuse the world; uncompelled by ways that use the world; characterized by self-control, by self-restraint, and by the ability to say "no" or "enough." Asceticism, then, aims at refinement, not detachment or destruction. Its goal is moderation, not repression. Its content is positive, not negative: It looks to service, not selfishness; to reconciliation, not renunciation or escape. Without asceticism, none of us is authentically human.

Let us, then, examine the significance of the ascetic practice of fasting. We Orthodox fast from all dairy and meat products for half of the entire year, almost as if in an effort to reconcile one half of the year with the other, secular time with the time of the kingdom. To fast is:

- not to deny the world, but to affirm the world, together with the body, as well as the material creation
- to remember the hunger of others, identifying ourselves with—and not isolating ourselves from—the rest of the world
- to feel the hunger of creation itself for restoration and transfiguration
- to hunger for God, transforming the act of eating into nothing less than a sacrament
- to remember that we live not "by bread alone" (Matt. 4.4), that there is a spiritual dimension to our life
- to feast along with the entire world, for we Orthodox fast together, never alone or at whim

To fast is to acknowledge that all of this world, "the earth, is the Lord's, and all the fullness thereof" (Ps. 23.1). It is to affirm that the material creation is not under our control; it is not to be exploited selfishly, but is to be returned in thanks to God, restored in communion with God.

Therefore, to fast is to learn to give, and not simply to give up. It is not to deny, but in fact to offer, to learn to share, to connect with the natural world. It is beginning to break down barriers with my neighbor and my world, recognizing in others faces, icons; and in the earth the face itself of God. Anyone who does not love trees does not love people; anyone who does not love trees does not love God.

To fast, then, is to love; it is to see more clearly, to restore the primal vision of creation, the original beauty of the world. To fast is to move away from what we want, to what the world needs. It is to be liberated from greed, control, and compulsion. It is to free creation itself from fear and destruction. Fasting is to value everything for itself, and not simply for ourselves. It is to regain a sense of wonder, to be filled with a sense of goodness, of God-liness. It is to see all things in God, and God in all things. The discipline of fasting is the necessary corrective for our culture of wasting. Letting go is the critical balance for our controlling; communion is the alternative for our consumption; and sharing is the only appropriate healing of the scarring that we have left on the body of our world, as well as on humanity as the body of God.

This conference is a golden opportunity for all of us to recognize the unique role of every individual and of every organization, in order that we may respect the more vulnerable in this situation, and in order that we may be prepared to assume responsibility for an issue of critical significance and global urgency.

May we be inspired by grace and justice, guided by reason and responsibility, and filled with selflessness and compassion.

RELIGION, SCIENCE, AND THE ENVIRONMENT SYMPOSIA

Official Opening, Symposium I, Istanbul, Turkey, September 22, 1995

REVELATION AND THE ENVIRONMENT

With the blessing of God, we have assembled like one of the scenes from the Book of Revelation. We come from many cultures and peoples, tribes and tongues, nations and faiths. Yet we share a common concern for the future of the planet and the quality of human life and relationships throughout the earth. We are assembled on a kind of latter day ark, in itself a symbol that would have delighted St. John. As we begin our voyage, the ship faces us with our essential interdependence. Here, in microcosm, is a truth that we need to bring to our distress in any part of the ship: The passengers dining in first class will not for long escape the consequences.

We shall be voyaging to Patmos on the sea of possibility, the sea from which life itself emerged. It is a time of profound cultural change and, at this moment, we are mindful of the Spirit of God, which in the beginning moved upon the face of the waters and which continues to move on these waters. We have been drawn together by a memory. Nineteen hundred years ago on the island of Patmos, St. John received the Revelation, which forms the final book of the New Testament. This occasion, however, is not intended to be a simple anniversary or an academic conference about an ancient text. Revelation begins and ends with the good news of the *parousia*, of the coming and the presence of Christ. At the climax of the New Testament, there is no full stop, but only an opening to the work of the Holy Spirit in the future, and the promise of a new creation: a new heaven and a new earth; a new community in a holy city; a river of life and a tree with leaves for the healing of nations. It seemed appropriate to celebrate this anniversary with a conference about our common home, the natural environment. St. John's vision is of a united human family—of every nation and kindred singing a new song.

THE STORY OF SYMBOLS

Much of the Bible is addressed to those with ears to hear, but the Revelation granted to St. John is also addressed to those with eyes to see. The story is told in symbols and archetypes. The root of the English word "symbol" lies in the Greek notion of bringing together fragments of truth in order to achieve a more profound understanding than would otherwise be available through mere analysis. Symbols help us to comprehend the relations between our sometimes fragmentary and fugitive perceptions of reality. Great symbols are not devised to illustrate some thesis that we may wish to advance. They arise rather from some deep level of consciousness and are disclosed to our reason. They have the power to communicate in a way that itself generates energy. Two such symbols have been entrusted to our generation. These are the "cloud" and the "globe."

In this year of anniversaries, we are all deeply aware of the mushroom cloud, which during the sacred Feast of Transfiguration, on August 6, 1945, opened a radically new chapter in human history. The cloud, however, is a symbol also found in the Bible. And, while the mushroom cloud is reflective of man's transmutation, it should not be understood in an entirely negative way. For the first time in history, by unraveling some of the forces that lie at the heart of creation, we have acquired the power to destroy all human life on the planet. By the same act, the world community is under threat. The work that lies ahead for all those who love life is to translate this world community, which exists as an object under threat, more and more into a subject of promise and hope.

In practice, of course, we extrapolate and anticipate in a single act, which is one of the reasons why science and theology need to be in dialogue at this particular moment of transition. Symbols of the End-Time combine with what we consider possible to create a field of action and to fill it with either hope or despair. Science saves faith from fantasy. Faith generates the energy for a new world. We face a need to communicate in ways that release healthful energies for restraint and change. Moralizing exhortations about the common good are limited in their usefulness. In the words of Carl Jung: "Mere appeals to ethical fraternity will never evoke in man that age-old power, which drives the migrating birds across the ocean."[8]

We hope and expect that this conference will increase our understanding of the various ways in which we may perceive and engage with the world around us. Together, as Orthodox Patriarchs, Metropolitans, and Archbishops, we seek your expert counsel, suggestions, and input so that Orthodox worldwide can better contribute to the common front being forged by intrepid scientists, environmentalists, and theologians who desire not only a pollution-free world, but a "healing of nations" as well. Indeed, our common future depends on developing a way to perceive and participate in the world, which will complement the analytical approach with an ecological awareness of entities in their various relationships.

The statement entitled "Orthodoxy and the Ecological Crisis," published under the aegis of our predecessor of blessed memory, Patriarch Dimitrios, reaffirmed that the monastic and ascetic traditions of the Orthodox Church have important insights for us. These traditions develop a sensitivity toward the suffering and beauty of all creation. "Love all God's creation," urged Dostoevsky, "the whole of it and every grain of sand. Love every leaf and every ray of God's light. Love the animals, love the plants, love everything. If you love everything you will perceive the divine mystery in things."

THEOSIS, METANOIA, AND ENKRATEIA

It is a life-creating tradition that beckons all to become a new creation in Christ, by being born of "water and spirit" (Jn 3.5), so that all matter, all life, may become sanctified. In order for sanctification or *theosis* (lit., deification) to become real, there must be a *metanoia*, a conversion or changing of the mind, reflective of the sanctity of tears. It is not a mere poetic coincidence that a contemporary Christian poet describes the rivers, seas, and oceans as "a gathering of tears" bearing witness to humanity's adventure and struggling journey. So, too, the Fathers of the desert

---

8. In a letter from C. G. Jung to Sigmund Freud, dated February 11, 1910. See *The Freud/Jung Letters: The Correspondence Between Sigmund Freud and C. G. Jung* (Princeton, NJ: Princeton University Press, 1974).

considered "the baptism of tears" as a lofty blessing empowering all men and women who seek "to come to the knowledge of the truth" (1 Tim. 2.4). Therefore, instead of asking for wisdom and strength and holiness, the angels of the desert asked for tears of repentance in their sojourn and struggle for salvation.

The ascetic tradition also offers a celebratory use of the resources of creation in a spirit of *enkrateia* (abstinence, but literally, "a holding within") and liberation from the passions. Within this tradition, many human beings have experienced the joy of contemplation, which contrasts with the necessarily fleeting and illusory pleasure of relating to the world as an object for consumption.

At the same time, we would want to celebrate the resources offered by the international community of scientists and journalists. As we enter the Mediterranean, it is encouraging to recall the achievements of the Med Plan, detailed by Peter Haas.[9] It is easy to be cynical about the impact of scientific experts and communicators on the calculations of nation states, but the Med Plan and its development ought to reinforce the urgency with which we seek to build up a network of personal relationships at even higher levels. We hope that this conference and its aftermath will serve as a major contributor to this growing family.

The Orthodox Church is particularly well represented in parts of the world where "the earth has been hurt," to adopt a phrase from the Book of Revelation. In a perversion of science, would-be God-slayers have laid waste to great tracts of territory. In these countries, the experience of the martyrs, which St. John also aptly describes, is very contemporary. We pray that the energy, which comes from giving up life for the sake of God and His Church, may flow into a life-giving stream for the benefit of the entire community of humankind.

Another symbol has been given to us, which points in this hopeful direction, although being more recent it has sunk more profoundly into human consciousness. The "earthrise" photograph taken in 1969 from the *Apollo* spacecraft shows the entire planet sapphire blue and beautiful, just as no human being since the dawn of history had ever seen it. This angelic view is foreshadowed and enlarged in the Book of Revelation,

---

9. In his book, *Saving the Mediterranean: The Politics of International Environmental Cooperation* (New York: Columbia University Press, 1990).

when John is shown the heavens opened and contemplates a great multitude out of every nation, rejoicing before the throne of God.

It may be that the choice between life and death always being put before us by the Spirit is in our day being translated into a choice between one world or more. Theology and science ought to be partners in this work. We ought not divide the one reality, but rather seek—as one theologian has expressed it—"peaceful co-existence [at the price] of mutual irrelevance."[10]

THE REVELATION OF HOPE

We recognize, however, that many people may have doubts about the possibility of traffic between the worldview expressed in modern science and the visionary material in the ancient Book of Revelation. How can Revelation's vision of hope, sustained in the midst of passages portraying terrible destruction, be distinguished from a rather unconvincing whistling in the dark to keep the spirits up in a time of danger and change? One approach is to consider the various ways in which our languages enable us to contemplate the future in as many languages as possible. We must contrast between the future—which is entirely constructed out of past and present, and which will itself soon become past—and the future which, beyond our control, will come upon us. In Greek, *ta mellonta* is what will be; whereas *parousia*, a term that is frequently used in the New Testament, suggests a future coming.

There is a similar distinction in Latin languages, derived from this same notion, between *futurus* and *adventus*. *Futurus* suggests a future entirely constructed out of past and present and, as such, it is a stimulus to planning. Futurologists of this school are those who rely on extrapolations from present trends. Often, of course, the collision between trends makes exact predictions. However, there is another problem with this approach to the future. Prolonging and projecting the present endorses present patterns of power and ownership and suppresses alternative possibilities, which the future holds. *Parousia*, on the other hand, alerts us to what is on its way to the present. In the Book of Revelation, Jesus Christ is hailed

---

10. See J. Moltmann, *God in Creation: An Ecological Doctrine of Creation* (London: SCM Press, 1985), 33.

as "the one who is and who was and who is to come" when the series might logically have been completed by Him who will be. A sense of the future as *parousia* stimulates the anticipation by which we attune ourselves to something ahead, whether through fear or through hope. These foretastes, symbolic sketches, and attainments, are part of every perception of the unknown, which we explore by reference to ultimate criteria such as happiness or unhappiness, as well as life or death.

The climax of our celebrations will be the Divine Liturgy on the island of Patmos. The liturgy is a work that binds human beings together in the Spirit of our Lord Jesus Christ and liberates them to hope and to work for His future coming in the world. Not everyone in this symposium shares the Christian faith. Nevertheless, we trust that we are all here around a common table like the symposiasts of ancient times because we all know that we are on the threshold of a new day. Conscious of the threat of nuclear destruction and environmental pollution, we shall move toward one world or none. We hope that we are assembled as those who are weary of defining their own tradition principally by excluding others. For many generations, the Patriarch of Constantinople has occupied what is known as the Ecumenical Throne. There is in that title a reaching forward to the "end-time" of the healing of the nations, when there will be communion between God and human beings in a new heaven and a new earth "and there shall be no curse anymore." We pray that this conference may contribute toward mobilizing all people of good will in our world and bringing closer together the day foreseen in the vision of Patmos.

Meditation Before the Cross, Symposium III, Passau, Germany, October 17, 1999

CREATION AND THE CROSS

Standing before the Cross, we are called to enter deeply into its great mystery: "For the word of the cross is folly to those who are perishing, but to us who are being saved it is the power of God" (1 Cor. 1.18). However, crucified power, power given to those who wanted to annihilate it, is a scheme not easily acceptable to human logic, which is accustomed to recognizing as stronger the one who avoids personal humiliation. However, in this case the grandeur and power of God is revealed in that it remains truly invulnerable to those who attempt to destroy it, in spite of the fact that it appears to be annihilated through the death of the divine Word on the Cross. The *Logos* remains untouched by death and by every threat. In this way, the Cross, which was formerly a symbol of defeat and shame, is now turned into a symbol of power and glory because the Crucified one is untouchable by the Cross. In other words, the Cross is proven powerless to touch the Word of God. Christ was subject to the greatest trial and was found stronger than it. To Him be the glory and honor.

In Christian societies, the Cross is venerated as a symbol of the voluntary passion of the divine Word and His victory over death. Although He who first carried the Cross calls those who want to follow Him to carry their own cross, only few joyfully accept to undergo this trial. For, the Cross is equated with the death of the ego and yet also raises the transfigured ego through its sacred identification, out of love, with every "thou" of the human race and the heavenly communion of personal spirits.

Nevertheless, the road of life passes through voluntary crucifixion and the road of eternity through the acceptance in time of death. The mystery of the Cross, while difficult to fathom and to accept, therefore stands before us demanding. It provokes the conscience and enters it as a living experience and as a challenge, not convinced beforehand by persuasive human wise words, but moving as strong as death, in faith and love (Sg. 8.6) when the demonstration in the spirit and power of God will come (1 Cor. 2.4).

Before each noble high achievement, there is sacrifice. And the symbol of such sacrifice is the Cross. If we want our efforts to succeed, as we begin our study and prayer of the Danubian environment from this point, in order to make the Danube a river of life, each of us must undergo some sacrifice. This sacrifice is the demonstration of our strength and brings upon itself the blessing and the synergy of God, which can overthrow every obstacle and achieve that which we are unable to characterize in any other way than as a miracle. Departing, now, for the miracle of Christ's resurrection, a symbol of our own resurrection and of the restoration of all nature, we stand uprightly before the Cross of Christ, in the power of which we wish to triumph.

Concluding Ceremony, Symposium IV, Venice, June 10, 2002

SACRIFICE: THE MISSING DIMENSION

As we come to the close of our fourth symposium on Religion, Science, and the Environment, we offer thanks to God for the fruitful proceedings as well as for your invaluable contribution. We recall the prophetic words of our predecessor, Ecumenical Patriarch Dimitrios I of blessed memory. In his historic encyclical letter of 1989, urging Christians to observe September 1st as a day of prayer for the protection of the environment, he emphasized the need for all of us to display "a eucharistic and ascetic spirit."

Let us reflect on these two words "eucharistic" and "ascetic." The implications of the first word are easy to appreciate. In calling for a "eucharistic spirit," Patriarch Dimitrios was reminding us that the created world is not simply our possession but it is a gift—a gift from God the Creator, a healing gift, a gift of wonder and beauty—and that our proper response, on receiving such a gift, is to accept it with gratitude and thanksgiving. This is surely the distinctive characteristic of ourselves as human beings: Humankind is not merely a logical or a political animal, but above all a eucharistic animal, capable of gratitude and endowed with the power to bless God for the gift of creation. Other animals express their gratefulness simply by being themselves, by living in the world in their own instinctive manner; but we human beings possess self-awareness, and so consciously and by deliberate choice we can thank God with eucharistic joy. *Without such thanksgiving, we are not truly human.*

A SPIRIT OF ASCETICISM

However, what does Patriarch Dimitrios mean by the second word, "ascetic"? When we speak of asceticism, we think of such things as fasting, vigils, and rigorous practices. That is indeed part of what is involved; but *askesis* signifies much more than this. It means that, in relation to the environment, we are to display what *The Philokalia* and other spiritual texts of the Orthodox Church call *enkrateia*, "self-restraint."

That is to say, we are to practice a voluntary self-limitation in our consumption of food and natural resources. Each of us is called to make the crucial distinction between what we *want* and what we *need*. Only through such self-denial, through our willingness sometimes to forgo and to say "no" or "enough" will we rediscover our true human place in the universe. The fundamental criterion for an environmental ethic is not individualistic or commercial. The acquisition of material goods cannot justify the self-centered desire to control the natural resources of the world. Greed and avarice render the world *opaque*, turning all things to dust and ashes. Generosity and unselfishness render the world *transparent*, turning all things into a sacrament of loving communion—communion between human beings with one another, communion between human beings and God.

This need for an ascetic spirit can be summed up in a single key word: sacrifice. This exactly is the missing dimension in our environmental ethos and ecological action. We are all painfully aware of the fundamental obstacle that confronts us in our work for the environment. It is precisely this: How are we to move from theory to action, from words to deeds. We do not lack technical scientific information about the nature of the present ecological crisis. We know, not simply *what* needs to be done, but also *how* to do it. Yet, despite all this information, unfortunately little is actually done. It is a long journey from the head to the heart, and an even longer journey from the heart to the hands.

REPENTANCE AND SACRIFICE

How shall we bridge this tragic gap between theory and practice, between ideas and actuality? There is only one way: through the missing dimension of sacrifice. We are thinking here of a sacrifice that is not cheap but costly: "I will not offer to the Lord my God that which costs me nothing" (2 Sam. 24.24). There will be an effective, transforming change in the environment if, and only if, we are prepared to make sacrifices that are radical, painful, and genuinely unselfish. If we sacrifice nothing, we shall achieve nothing. Needless to say, as regards both nations and individuals, so much more is demanded from the rich than from the poor. Nevertheless, all are asked to sacrifice something for the sake of their fellow humans.

Sacrifice is primarily a spiritual issue and less an economic one. In speaking about sacrifice, we are talking about an issue that is not technological but ethical. Indeed, environmental ethics is specifically a central theme of this present symposium. We often refer to an environmental crisis, but the real crisis lies not in the environment but in the human heart. The fundamental problem is to be found not outside but inside ourselves, not in the ecosystem but in the way we think.

The root causes of all our difficulties are human selfishness and human sin. What is asked of us is not greater technological skill but deeper repentance, *metanoia*, in the literal sense of the Greek word, which signifies "change of mind." The root cause of our environmental sin lies in our self-centeredness and in the mistaken order of values, which we inherit and accept without any critical evaluation. We need a new way of thinking about our own selves, about our relationship with the world and with God. Without this revolutionary "change of mind," all our conservation projects, however well intentioned, will remain ultimately ineffective. For, we shall be dealing only with the symptoms, not with their cause. Lectures and international conferences may help to awaken our conscience, but what is truly required is a baptism of tears.

THE BIBLICAL VIEW OF SACRIFICE

Speaking about sacrifice is unfashionable, and even unpopular, in the modern world. But, if the idea of sacrifice is unpopular, this is primarily because many people have a false notion of what sacrifice actually means. They imagine that sacrifice involves loss or death; they see sacrifice as somber or gloomy. Perhaps this is because, throughout the centuries, religious concepts have been used to introduce distinctions between those who have and those who have not, as well as to justify avarice, abuse, and arrogance.

Nevertheless, if we consider how sacrifice was understood in the Old Testament, we find that the Israelites had a very different view of its significance. To them, sacrifice meant not loss but gain, not death but life. Sacrifice was costly, but it brought about not diminution but fulfillment; it was a change not for the worse but for the better. Above all, for the Israelites, sacrifice signified not primarily giving up but simply giving. In its basic essence, a sacrifice is a gift—a voluntary offering in worship by humanity to God.

Thus in the Old Testament, although sacrifice often involved the slaying of an animal, the whole point was not the taking but the giving of life, not the death of the animal but the offering of the animal's life to God. Through this sacrificial offering, a bond was established between the human worshipper and God. The gift, once accepted by God, was consecrated, acting as a means of communion between Him and His people. For the Israelites, the fasts—and the sacrifices that went with them—were "seasons of joy and gladness, and cheerful festivals" (Zech. 8.19).

An essential element of any sacrifice is that it should be willing and voluntary. That which is extracted from us by force and violence, against our will, is not a sacrifice. Only what we offer in freedom and in love is truly a sacrifice. There is no sacrifice without love. When we surrender something unwillingly, we suffer loss; but when we offer something voluntarily, out of love, we only gain.

When, on the fortieth day after Christ's birth, His mother the Virgin Mary, accompanied by Joseph, came to the temple and offered her child to God, her act of sacrifice brought her not sorrow but joy, for it was an act of love. She did not lose her child, but He became her own in a way that He could never otherwise have been. Christ proclaimed this seemingly contradictory mystery when He taught: "Whosoever wishes to save his life must lose it" (Matt. 10.39 and 16.25). When we sacrifice our life and share our wealth, we gain life in abundance and enrich the entire world. Such is the experience of humankind over the ages: *Kenosis* means *plerosis*; voluntary self-emptying brings self-fulfillment.

All this we need to apply to our work for the environment. There can be no salvation for the world, no healing, no hope of a better future, without the missing dimension of sacrifice. Without a sacrifice that is costly and uncompromising, we shall never be able to act as priests of the creation in order to reverse the descending spiral of ecological degradation.

THE LITURGICAL VIEW OF SACRIFICE

The path that lies before us, as we continue on our spiritual voyage of ecological exploration, is strikingly indicated in the ceremony of the Great Blessing of the Waters, performed in the Orthodox Church on January 6, the Feast of Theophany, when we commemorate Christ's baptism in the Jordan River. The Great Blessing begins with a hymn of praise to God for the beauty and harmony of creation:

Great are You, O Lord, and marvelous are Your works: no words suffice to sing the praise of Your wonders. . . . The sun sings Your praises; the moon glorifies You; the stars supplicate before You; the light obeys You; the deeps are afraid at Your presence; the fountains are Your servants; You have stretched out the heavens like a curtain; You have established the earth upon the waters; You have walled about the sea with sand; You have poured forth the air that living things may breathe. . . .

Then, after this all-embracing cosmic doxology, there comes the culminating moment in the ceremony of blessing. The celebrant takes a cross and plunges it into the vessel of water (if the service is being performed indoors in church) or into the river or the sea (if the service takes place out of doors).

THE WAY OF THE CROSS

The Cross is our guiding symbol in the supreme sacrifice, to which we are all called. It sanctifies the waters and, through them, transforms the entire world. Who can forget the imposing symbol of the Cross in the splendid mosaic of the Basilica of Sant'Apollinare in Classe? As we celebrated the Divine Liturgy in Ravenna, our attention was focused on the Cross, which stood at the center of our heavenly vision, at the center of the natural beauty that surrounds it, and at the center of our celebration of heaven on earth.

Such is the model of our ecological endeavors. Such is the foundation of any environmental ethic. The Cross *must* be plunged into the waters. The Cross *must* be at the very center of our vision. *Without the Cross, without sacrifice, there can be no blessing and no cosmic transfiguration.*

PUBLICATIONS AND REFLECTIONS

Pastoral Encyclical, Istanbul, Turkey, December 25, 1994

CHURCH AND SYMBOLS

The Christian Church was revealed—and is constantly being revealed—by the one God, the almighty Creator of heaven and earth and of all things visible and invisible. These revelations are evident through the human nativity of the consubstantial Word of God. The Church has crossed the narrow confines of Palestine and the Mosaic Law and has ventured out to encounter and dialogue with the world of the Gentiles. Since the beginning, the Church did not hesitate to embrace all that the divine Creator had made in His infinite providence and love. Through the life of doxological worship, particularly through the supreme expression of the divine Eucharist, the very emblems of the Gentile divinities were returned to their natural purity and a new symbolism emerged: *the Christocentric reality of the cosmos and nature.*

In the catacombs of ancient Rome, Christ was portrayed as a pure white lamb, as a fish, or as a vine. The four Evangelists were represented by an angel, a lion, an ox, and an eagle. The Holy Spirit has been depicted as a dove ever since the baptism of Jesus Christ in the River Jordan. With the passage of time, the Church became even more audacious in making use of nature's symbols in order to decorate the space where the faithful worshipped. The church interior, in fact liturgy itself, became a miniature icon of the universe, of heaven, of earth, of the netherworld, and of the world to come. We can begin, therefore, to understand the concepts of sacred space and sacred time.

ENVIRONMENT AND SIN

While the plenitude of theological vision in Jesus Christ allows the highest doxological offering of the universe to the almighty, the thoughtless and abusive treatment of even the smallest material and living creation of God must be considered a mortal sin. An insult toward the natural creation is seen as—and in fact actually is—an unforgivable insult to the uncreated God.

At this particular juncture, the Christian Church, and especially the Orthodox Church, turns its attention toward the land of the Rising Sun and the delicate sensitivity of the spiritual vision of nature found in Buddhism, which has shaped the consciences and souls of the noble Japanese race. It is extremely significant that the Church observes in Japanese life, particularly in its expression of art, an overwhelming preponderance of the beauty and grace of God's creation and the profound respect for it. So often, there is a hardly susceptible mysterious element of intuitiveness of the temporality of the subjects portrayed. Indeed, the realization that the entire visible world has a finite existence, that "it fades like a flower," is for all of us the beginning of the most existential inner searching about what succeeds death.

Orthodoxy has its response: "resurrection from the dead," "new creation," and "a better, enduring existence" (Heb. 10.34). For, as St. Paul emphasizes, "our citizenship is in heaven" (Phil. 3.20). Nevertheless, the mystery of creation can be appreciated only in faith, without which true knowledge of God would be absolutely unobtainable. This faith has been delivered given "once for all" (Jude 3) as creation itself was created once for all.

Foreword to the English Translation of Vespers for the Protection of the Environment, *2001*

PRAYING FOR THE EARTH

God is blessed because He is Love.[11] When we live love we broaden our existence; it becomes something without end: "For love never ends" (1 Cor. 13.8). There can never be too much love; there is no over-saturation of love nor a turning away from love. He who loves, gathers all things together under the shelter of that love. If a sinner, or self-centered person can feel even a limited love, then surely he is also able to comprehend how much greater that happiness would be if he broadened that love which brought him so much joy.

God, being the very essence of love, could not as an entity be single-faceted because love is a feeling directed to another entity of identical essence. Literally, it is a feeling expressed only between persons. Thus, the very nature of God as love unavoidably leads to the begetting of the Son, by the Father, before all ages, and also the emanation of the Holy Spirit. These are personal hypostases able to love the Father and one another, and be loved by the Father.

Love turned inward to one's self, without loving another is to be sure, not love, and does not in such an instance therefore, give the true, precise meaning of the word. The love of God the Father, and the other two

---

11. Composed by Monk Gerasimos of the Community of Little St. Anne on Mt. Athos and translated by Rev. S. Kezios of California. The foreword was prepared in June, 1997, but published in 2001. An earlier translation was published by Archimandrite Ephrem (Lash).

Persons of the Holy Trinity with Him, could not be limited to the limitless bounds of love between them. In this particular instance this does not mean that He would be a God dependent on others (for this would be blasphemy against God who in the fullness of His blessedness is self-sufficient).

Expressed in terms of human understanding, the meaning is that the overflowing of His love could not possibly take on a form of existence unless directed toward something capable of accepting this overflowing of His love. The quintessence of this overflowing love of God, capable of giving and receiving love, was the boundlessly abundant, personal, spiritual and body-spirit creation of angels and people, and the infinite creation of the whole universe to serve them.

The angels, spiritual beings composed of a spiritual essence different from that of God, are in a more immediate relationship with Him. Their exultation is expressed through the wondrous and incomprehensible majesty of their life, in that they are loved by the Triune God. In turn, they love Him by their unending glorification of Him and their immediate conformance to His will.

GOD'S LOVE FOR THE WORLD

Somewhat lower than angels, people were created by God in His image and likeness and were crowned by God with glory and honor. As such, they are capable of both receiving and returning His love, resulting in their true happiness. Indeed, the breadth of God's love for us is manifested in terms of the limitless number of persons we are able to love and in turn be loved. People experience a taste of this love (the breadth of God's love) when, irrespective of their religious credo, they realize the joy of loving and being loved by many.

God's love for humanity is unsurpassed and cannot be equally reciprocated because humanity turned its face from Him and from His love, whereas God continues to love us. This aforementioned love is even more unrivaled because God expressed the infinite richness of His love by the creation of a material world of incomprehensible beauty, variety, and expanse. This was done for the sake of man who comprises both matter and spirit. When one considers all that is offered to man for his enjoyment in sight, sound, taste, smell, and touch, it is more than enough to put us in awe before the benevolence of God.

The first-created human being was placed by God in a beautiful garden on planet Earth. But then again, the whole of Earth was adorned by so many such splendid things that the psalmist, citing but a few, concludes in total wonder, "How great are your works, O Lord! In wisdom you have wrought them all" (Ps. 103 [104]). Even more so, it is to be found in the infinite richness of the visible microcosm, discovered through the aid of scientific instruments, and also, in the grandeur of the indescribable celestial universe whose dimensions we measure in millions of light years.

If anyone says, "I love God," and hates one's brother . . . that person cannot love God (1 Jn 4.20). Anyone, however, who loves one's neighbor and God, becomes resplendent because of this love:

> . . . for all creation, for people, and birds, and animals, and demons, and for every creature . . . and one cannot bear or hear, or behold any harm whatever or any minor affliction occurring in nature. For this reason, this person offers a tearful prayer in every hour of every day for the animals (irrational beings), for the enemies of truth, and for those who would harm him, that they may be protected and receive expiation; and out of the great, immeasurable mercy, motivating the heart in imitation of God, prays as well for the reptilian species.[12]

If these feelings toward the whole of creation are nurtured by a monk who has abandoned the world and its cares, then what should a person who lives in the world feel as a sense of duty to creation? Obviously, one's first sentiment should be to give thanks and glorify God who granted us the whole of creation for our use and enjoyment. This is especially so for that which is more proximate and which we call the environment. St. Paul gives us this charge when he writes: "Pray constantly, give thanks in all circumstances" (1 Thess. 5.17–18).

In fulfillment of this charge, the Holy Great Mother Church of Christ has designated the first day of September as a world day of prayer for the environment. That is to say, for the creation granted by God to humankind, which instead of being protected, as commanded by God (Gen. 2.15), is often abused.

---

12. *Ascetic Treatises*, Homily 84.

*Reflection on Theme, General Assembly of the World Council of
Churches, São Paulo, January 2006*

A CALL TO TRANSFORMATION

The Divine Liturgy of St. John Chrysostom is the eucharistic service that
has been celebrated by Orthodox Christians throughout the world since
the fifth century, when it was attributed to the remarkable preacher and
renowned Archbishop of Constantinople. The central words of the liturgy
are proclaimed by the celebrant praying on behalf of and with the entire
community: "Send down your Holy Spirit upon us and upon these gifts,"
namely the bread and wine that symbolize the life of the world. The
celebrant then continues: "Transform them into your sacred body and
precious blood." And the community responds with the unique liturgical
repetition: "Amen. Amen. Amen." This is followed by a profoundly re-
newing moment of sacred silence.

There are three crucial and essential dimensions in this liturgical expression of transformation:

Transformation begins within the human heart because the divine Spirit is
first invoked "upon us" as human beings.

Transformation occurs within the wider community because "the gifts" are
offered by and for the entire community.

Transformation occurs for the whole creation because bread and wine are
representative of the natural environment.

THE HEALING OF THE HEART

The early spiritual literature of the Christian East has, through the centuries, emphasized the heart as the place of transformation, where God,
humanity, and the world coincide and coexist in a relationship marked
by prayer and peace. *The Philokalia* underlines the astonishing paradox
that the transformation of all things is only achieved through inner silence: "When you find yourself in silence, then you will find God and
the world entire!" In other words, transformation begins with the awareness that God, and God alone, is to be held at the center of all life. The
grace of God is closer, more integral to us, indeed more definitive of us,
than our own selves! This is why St. Gregory Palamas defended the

"prayer of the heart" as a powerful way of realizing how "the kingdom of God is within us" (Lk 17.21).

While the ways of silence and serenity are nurtured in a unique way in the Christian East, they are of course neither a monopoly of the Orthodox Church nor of Christianity itself. The exhortation of the Old Testament Psalmist is: "Be still, and know God" (Ps. 44.1). Moreover, the Arabic root of the word Islam connotes a sense of holistic transformation, of wholeness and integrity. Ordering one's relationship to God, others, and the world is the Muslim state of "salam," which is closely related to the Hebrew word for peace, "shalom." How, then, is it possible to hear the Word of God—whether as Christians, Jews, or Muslims—unless we first stop to listen in silence? How can we ever be sure that we are working to transform the world around us unless we have first transformed the world within us?

Inner transformation, however, requires radical change. In religious terminology, it requires conversion or *metanoia*—a change in attitudes and assumptions. We cannot be transformed (or converted) unless we have first confronted everything that stands in opposition to transformation; we cannot be transfigured until we have been cleansed of everything that disfigures the human heart as it was created and intended by God. Such a process of self-discovery leads ultimately to the respect of human nature, with all of its flaws and failures—both in ourselves and in others. It paves the way for the respect of every human being, irrespective of differences—within society and within the global community. Indeed, these differences are to be welcomed, honored, and embraced as unique pieces of a sacred puzzle, the mystery of God's wonderful creation.

THE WAY OF COMMUNITY

The healing of the heart leads to the way of community. Transformation is a vision of connection and compassion. How unfortunate it is that we Christians have disassociated spirituality from community. When as Orthodox Christians we depart from the transforming event of the Divine Liturgy, we move out to the same world, the same routine, and the same problems. Yet, now, we can see otherwise; we now know differently; we are now impelled to act graciously. When we are transformed by divine grace, then we shall seek solutions to conflict through open exchange without resorting to oppression or domination.

We have it in our power either to increase the hurt inflicted in our world or else to contribute toward its healing. When will we realize the detrimental effects of war on our spiritual, social, cultural, and ecological environment? When will we recognize the obvious irrationality of military violence, national conflict, and racial intolerance, all of which betray a lack of imagination and willpower? Transformation involves awakening from indifference and extending our compassion to victims of war, poverty, and all forms of injustice. As faith communities and as religious leaders, we must proclaim alternative ways that reject war and violence and which recognize peace as the only way forward. Human conflict may well be inevitable, but war and violence are not. Human perfection may well be unattainable, but peace is not impossible. If this century will be remembered, it will be remembered for those who dedicated themselves to the cause of peace. We must believe in and "pursue what makes for peace" (Rom. 14.19).

Indeed, transformation is our only hope of breaking the vicious cycle of violence and injustice, and it is vicious precisely because it is the fruit of vice. War and peace are systems; they are contradictory ways of resolving problems and conflicts. Ultimately, they are choices. This means that making peace is a matter of individual and institutional choice, as well as of individual and institutional change. It, too, requires conversion or *metanoia*—a change in policies and practices. Peacemaking requires commitment, courage, and sacrifice. It demands of us a willingness to become communities of transformation and to pursue justice as the prerequisite for global transformation.

THE RENEWAL OF THE EARTH

Over the last two decades, the Ecumenical Patriarchate has made the preservation of the natural environment a central focus of its spiritual attention and a priority of its pastoral ministry. We consider the healing of the heart and the way of community as integrally linked with the survival of our planet as well as with the way its inhabitants relate to the natural creation. A responsible relationship between the soul and its Creator and among human beings inevitably involves a balanced relationship with the natural world. The way we treat each other is immediately reflected in the way we treat our planet; the way we respond to others is at once measured and mirrored in the way we respond to the air we breathe, the water we drink, and the food we consume. In turn, these influence and reflect the way we pray and the way we worship God.

Whenever we narrow religious life to our own concerns, then we overlook the prophetic calling of the Church to implore God and invoke the divine Spirit for the renewal of the whole polluted cosmos. For, the entire world is the space within which transformation is enacted. When we are transformed by divine grace, then we discern the injustice in which we are participants; then we labor to share the resources of our planet; then we realize that eco-justice is paramount—not simply for a better life, but for our very survival.

Like the healing of the heart and the making of peace, ecological awareness also requires conversion or *metanoia*—a change in habits and lifestyles. Paradoxically, we become more aware of the impact of our attitudes and actions on other people and on the natural environment when we are prepared to surrender something. This is why fasting is a critical aspect of Orthodox Christian discipline: In learning to give up, we gradually learn to give. In learning to sacrifice, we gradually learn to share. Unfortunately, so many of our efforts toward reconciliation—whether spiritual, social or ecological—prove fruitless partly because we are unwilling to forgo established ways as individuals or as institutions, refusing to relinquish either wasteful consumerism or prideful nationalism. A transformed worldview enables us to perceive the immediate and lasting impact of our practices on other people (and especially on the poor, our neighbor) as well as on the environment (our silent neighbor).

TRANSFORMATION AND PROMISE

In the spiritual classics of the Orthodox Church, transformation signifies a foretaste of the kingdom to come. It can never be fully realized or exhausted in this world; it always tends and extends toward the heavenly world, which informs and imbues this world with sacred meaning. Christians should remember that the Church is called not to conform to, but to transform this world. The ultimate goal is not compromise with this world, but the promise of another way of seeing, living, and acting.

Such is the conviction of the Orthodox Easter liturgy, when the Resurrection of Christ is proclaimed as "the first-fruits of another way of life," "the pledge of a new beginning." Transformed in the light of Mt. Tabor and the Tomb of Christ, we can see the same things differently; we can march to a different drum—sometimes inevitably clashing with established patterns, with unquestioned practices, and accepted norms.

Transformed in this way, Christians become a grain of mustard seed, a form of leaven. They become enthusiastic and joyful witnesses to the light of the kingdom in our world. And there is only one way that we shall, with the grace of God, prevail as people and communities of transformation: together! Individuals and institutions are easily exhausted and discouraged if they act in isolation. The vision of the Psalmist is within our grasp: "Behold, it is a good and pleasant thing for us to dwell together in unity" (Ps. 133.1). Such is the imperative of the ecumenical vision of transformation.

6

Education and Economy

Conservation, Education, and People

Opening Ceremony, First Summer Seminar, Halki, Turkey,
June 20, 1994

ENVIRONMENT AND RELIGIOUS EDUCATION

By the grace of God, to whom we express our heartfelt gratitude, we are convening this interfaith gathering in this venerable center of Orthodoxy. We are pleased that it is being held within the hospitable environment of this monastery, where our "alma mater," the Theological School of Halki, once flourished and this year celebrates the 150th anniversary since its establishment.

Human beings were created by God to enjoy sovereignty over nature but not to exercise tyranny over it. Many sectors of society have now recognized that the ecological problem is associated with the moral crisis of the humanity and that the proper use of nature depends on the perception, position, and practice of human beings in relation to the cosmos. The ancient saying holds true: "Humanity is the measure of all things."[1]

Now that we have been awakened to the impending destruction of nature, and to all that this implies, how has society responded in recent years? We note here the so-called "plans for peaceful coexistence" (between humankind and nature) along with plans for the "development of

---

1. An axiom of the ancient Greek sophist, Protagoras, in the fifth century BC.

the environment." All these concerns and actions are, of course, commendable. As we know, however, they are also limited in their effectiveness. For, who will find and apprehend those individuals responsible for forest fires? Who will restrain those who illegally cut down trees? Who and how will we control those unconscionable individuals who pollute our waters, rivers, and seas? Who can restrain the greedy?

We, the Church, must assist willingly, firmly, and extensively, with this pressing and vital concern. We can help by enlightening the conscience of men and women in order to cultivate respect for their fellow human beings and for all created matter. Our goal is to instill in people a sense of feeling, as well as a sense of the fear of God so that they may avoid wrongdoing, vulgarity, impropriety, inhumanity, and especially selfish individualism. Usually those who torch forests, those who illegally cut down trees, and those who pollute our shores are egocentric individuals with hardened hearts, who do so out of greed and for purely utilitarian purposes. A good Christian cannot, rather a good Christian is by conscience not permitted to, destroy nature and the environment. A good Christian cannot be a source of immoral or destructive acts.

According to Socrates, "virtue is taught." In conformity with our position, therefore, and following much meditation and thought, we have chosen as the theme of this gathering the relationship between religious education and the natural environment. By restricting the discussion of the conference to religious education, we are neither excluding nor underestimating other forms and levels of secular and parochial education. Our goal—and we beseech the attention of the esteemed participants in this point—is to examine and explore ways and means by which we may sensitize and influence the desires and the attention, especially of our students, to this most urgent issue.

PARISH EDUCATION AND COMMUNAL ACTION

However, in order that we are not left with empty or vain words, we are of the opinion that our attention must be given to developing programs of practical application. For example, tree-planting initiatives must be undertaken, just as we have done today, and as we did last December on this island. Groups of students may cultivate gardens, while others can care and tend to forest regions. Along with a series of lectures, seminars should be organized with the express purpose of enlightening students on planting procedures, gardening, and other similar activities. Other groups of children in our secular, parochial, and catechetical schools may adopt vegetable or flower gardens, forested regions, church compounds, abandoned properties, or farm regions cultivated for the common good, as well as areas with natural beauty, which they will care for on a voluntary basis. Their example can serve to sensitize their parents and elders who can then be motivated to do likewise.

We especially advise the clergy and others in parish ministry to encourage and promote a love for nature, to care for trees and shrubs as well as church properties and cemeteries. It is only fitting that love for the environment begins in the church compound, which must be replete with greenery and flowers in bloom throughout all seasons of the year, "for the creator of beauty has made them all" (Wisd. of Sol. 13.3).

Such beauty of nature is the will of God. Consequently, we are obligated to preserve rather than to destroy the environment. Hence, any destruction of nature clearly constitutes sin. We entreat your attention particularly to these final thoughts. As we read the agonizing warnings of conservationists, geologists, biologists, and other specialists, who remind us of the great folly of the violation of nature with its foreseeable tragic consequences, you, my beloved speakers and participants, are today contributing to a momentous and monumental task of timely significance for our planet.

The prophet Isaiah declared: "God did not create the earth as chaos; rather, He formed it to be inhabited" (Is. 45.18). Humanity is obligated, therefore, not to ravage the earth, creating chaotic conditions with fires and scarcity of water, but rather to develop and enhance it. As a modern songwriter puts it: "You who have nothing to do, plant a tree in the corner of your garden so that others may come and sit there to rest and recollect." These are the words of Salvatore Adamo, in a timely song with beautiful orchestral rhythm and harmony. It would be worthwhile for our youthful listeners to find it. Inspired and incited by it, sing it along with your friends as an indication of your ecological concern. In life, only those divinely and enthusiastically inspired, only those who truly love their environment, are able to create the things of God. In the words of the author of the first book of Scripture: "Seedtime and harvest, cold and heat, summer and winter, day and night, shall not cease" (Gen. 8.22).

You, our beloved children in the Lord, are contributing to the concern for a proper order and legitimate status of the cosmos. Cosmos means decoration; it is defined as a love for beauty and decency. May you, therefore, be blessed by God.

THE LEAST PERSON—THE GREATEST DIFFERENCE

Permit us to confide in you our thoughts. We do not place much trust in the strong and the mighty, or in people of authority. We believe, rather, in those willing and patient individuals, in those who do not lose sight of their objective, namely the objective for good. Do not forget the acknowledgment of the ancient Greeks that "drops of water can make even rocks hollow." Many simple people, in various small corners of the earth, with nominal but continual daily concerns, are able to change the world, even if only slightly, for the better. Today, on the day after Pentecost, we celebrate the Feast of the Holy Spirit. We celebrate the triumph of the few, the weak—at least by human standards—holy disciples and Apostles of our Lord, who, empowered by the fire of Pentecost, changed the world some 2000 years ago, for all time, so that today we are preparing to enter the third millennium after Christ. My brothers and sisters, may we continue primarily to cultivate the field of the soul, but also the garden of our home, so that future generations may reap the fruit of our labor.

Foreword to Proceedings of the First Summer Seminar, *June 2000*

RELIGIOUS EDUCATION AND RELIGIOUS OBLIGATION

God, who created the world and humankind, placed those whom he first created in a specially selected garden on the terrestrial globe, according to the teaching of our faith. And He commanded them: "to labor in it and to tend it."

The one aspect of this two-part composite commandment establishes human labor as an obligation ordained by God, giving humanity the capacity of creating while laboring and, thus, of participating to a degree in the creative force of God. This constitutes one of the manifestations of human nature as fashioned in the image of God.

However, according to the second aspect of this commandment, what is recapitulated in the tending of the Garden in Eden is the religious and theological basis for the protection of the environment. Tending the garden certainly did not imply protecting it from any assault by some third party (since then there were no such!). Rather, it meant safeguarding it and maintaining it in the same condition as it was given by God to Adam, certainly for his use.

Later, having conformed to the divine commandment to increase and have dominion over the earth, when people in fact did disperse over the entire surface of the planet Earth, the commandment to safeguard the earth properly as their habitation understandably spread beyond the confines of the Garden of Eden throughout the entire earth.

Thus, today, we are able to say that the Christian religion, the Jewish religion, and the Muslim religion—the latter of which accepts in part the Old Testament as encompassing the declaration of God's will to humankind—are obliged to emphasize to their faithful that tending the earthly environment and, in general, the worldly environment that we inhabit, is a commandment of God. As such, it is a religious obligation.

Foreword to Proceedings of the Third Summer Seminar, *June 2000*

COMMUNICATION, COMMUNION, AND ENVIRONMENT

Within creation, all beings, animate and inanimate, communicate with one another and endure the same natural consequences, one influencing the other. Nevertheless, the human person is the supremely rational being of communication, endowed with the consciousness of dialogue with one's fellow human beings. This potential of conscientious and rational communication renders possible also the transmission of knowledge, ideas, and sentiments, thereby shaping humanity's stance before the rest of creation. The transmission today of distorted forms of communication, which look to the transformation of the human person into an individual easily manipulated by a formless and colorless social mass, surely constitutes a phenomenon for concern.

Communication or dialogue is a fundamental characteristic of life in the Church. The interpenetration of the three persons of the Holy Trinity constitutes the ultimate and perfect form of communication. The Church is a body of persons in continual communication with one another. The Holy Eucharist itself, that supreme sacrament of the Church, is the central means of communication between God and humanity in Christ. More particularly, the offering of creation by humanity to God in an act of thanksgiving constitutes the expression of the greatest ecological ethos. It is not by accident that the terms used to describe this great mystery are "divine Eucharist" and "divine communion."

The third Summer Ecological Seminar gathered renowned leaders in the areas of the Church, science, and scholarship as well as communications in order for these, together, to contribute to a more universal confrontation of the worldwide ecological crisis. Their papers and the recommendations of the groups and workshops serve as a rich resource for the tireless environmental efforts of the Patriarchate but also of every religion and person that is concerned about the developments toward the destruction of creation.

Official Reception, Symposium III, Galati, Romania,
October 25, 1999

CONCERN FOR NATURE—CARE FOR PEOPLE

It is with great joy that we arrive in the famous city of Galati, whose patron saint and protector is the Apostle Andrew. We feel that we are in familiar and fraternal surroundings, inasmuch as you lie under the protective care of the "first-called" of the Apostles, who is also under the paternal protector of our Mother Church of Constantinople. We are sincerely moved and feel deeply grateful for this honor and reception, as well as for the kind words, which confirm that we are of one soul with each other, and that we share a very close and unbreakable bond of faith and love that already unites our sister Churches.

As known, we are participating in and initiating the third international symposium, Religion, Science, and Environment, which this year is dealing with the theme "Danube—A River of Life." This subject is of direct concern to all Danubian peoples, but also to the peoples of the Black Sea region. For the Black Sea is a receptor of the waters from the Danube and of all the substances transferred from there. Our participation in this symposium is a result of our paternal interest for the welfare and peace of humanity, which largely depends on the plenitude of natural resources, and which very much depends also on the preservation of the natural and ecological health and balance. We regard it as the obligation of all Christians to show respect toward our fellow neighbor, whose life would otherwise be negatively affected by pollutants and by our polluting actions. This is why we preach this message of love and solidarity, and of respect for nature and humanity. The aim is for all of us to be sensitized in regard to this problem, and to become conscious to its resolution, each in our own field, from our particular position and captivity.

Opening Ceremony, Third Summer Seminar, Halki, Turkey, July 1, 1996

ENVIRONMENT AND COMMUNICATION

We do not pretend that we have sufficiently cultivated the ground upon which Churches, religious communities, environmental organizations, and scientists can communicate with each other and to the four corners of the earth about the vital issues of environmental protection. On the contrary, we steadfastly believe that, as we discover more and more the truly unbelievable dimensions of the worldwide ecological problem, a long-term program of complex and concerted studies is required. In the combined, creative synthesis of their conclusions, we may be able to face more effectively the global threats, caused by the irresponsible, if not criminal, behavior of rational human beings toward the non-rational and immaterial creation. In this creation, God has placed humanity from the beginning "in a paradise of delight," not only as sovereign, but also as healer and steward. Our destructive management of creation, besides having a practical impact on the quality of life, assumes a critical, moral dimension that constitutes a profound disrespect toward the Creator.

Moreover, the subject of the environment in general, which is being addressed here today, together with the much discussed issue of communication, is certain to open up very important perspectives for the tasks of this present seminar—for which we rejoice personally and most warmly in the name of the Mother Church.

It is clear, that through these words of greeting for the opening of your deliberations, in no way do we intend to prejudice or even to influence substantially the work that has been planned because indeed we are not speaking here as one of the experts on the subject. It is rather our desire to express a few general prefatory statements regarding the authoritative position of the Church on the whole subject of environment, especially in relation to the development of increased communication, which is of eminently vital significance for us, for the benefit of the whole of humankind and, ultimately, for the glory of God.

In short, it is sufficient to state that, just as it is important for the various systems of the human body (such as the nervous, digestive, and circulatory systems) to communicate with each other in order to maintain good health, the free operation of communication has the same value and purpose. Hence, the entire physical network of varying biotopes and specific geographical ecosystems will be better served not only through the mutual exchange of information but also by the coordination of activities among all those responsible.

COMMUNION AND COMMUNICATION

Let us express this same truth in its most spiritual form, and in a manner greatly pleasing to God. Communion with God in prayer, and solidarity and interaction with one's fellow human beings in every real situation, render truly blessed the distribution of all the good things of this present world. In the same way, unhindered communication among all those concerned with the management of the ecological realities of the present time is equivalent to the indispensability and the sanctity of prayer. Ultimately, all things created may be connected eucharistically for the praise of the one Creator and Father God.

With all our soul, and in the true spirit of prayer and service for the whole of humanity along with the whole of creation, we wish that the blessed efforts and tasks of Summer Seminar on Halki '96 may bear fruit. We profoundly thank first our respected governmental authorities for the assistance they have provided. Then, we must thank his Royal Highness, Prince Philip, Duke of Edinburgh who, once more, was well pleased to allow us to place this seminar under our joint aegis. Moreover, we thank all those who sent messages of goodwill for this seminar—in particular His Excellency, the Hon. Bill Clinton, President of the United States, who donated on behalf of the American Nation the young tree that we shall plant in a short while as a token of his esteem for the ecological endeavors of the Ecumenical Patriarchate. The expressions of support by these esteemed people demonstrate their strong interest in global environmental problems, and indicate humankind's increasing awareness of looming environmental threats. In this, too, communication has played a vital role.

This support also encourages the Ecumenical Patriarchate to continue in its undiminished and persistent initiatives to mobilize the moral and spiritual forces of the Orthodox Church in order to realize once more the harmony that existed between humanity and nature, to the glory of the Creator. This is because the Ecumenical Patriarchate continues to observe the rapid deterioration of environmental conditions globally, which frequently results in irreversible changes. However, the Ecumenical Patriarchate sees two hopeful phenomena. On the one hand, it recognizes the adoption by the international community of the principles of sustainability in the management of natural resources and a wiser conception of the development process; on the other hand, it appreciates the growing mobilization of people, and especially of the younger generation, in combating threats and managing the planet in a more sensible way.

Opening Ceremony, Fifth Summer Seminar, Halki, Turkey,
June 14, 1998

ENVIRONMENT AND POVERTY

It has been already four years since the Holy and Great Mother Church of Christ took up the initiative to hold a series of summer ecological seminars in this ancient and holy monastery. The fifth of these seminars commences today. Although five successive seminars are not sufficient to characterize their realization as an established institution, we are able to say that the timeliness and the acuteness of the ecological problem, whose solution cannot be immediately predicted, demand an increase and not a reduction of our efforts. Thus, what is needed is to expand and not to curtail the offerings of these seminars, which, God willing, would consequently result in their consolidation.

As is known, the first ecological seminar that took place here in 1994, was entitled "Environment and Religious Education." The following year, 1995, the seminar was concerned with the topic Environment and Ethics. The presentations and reports of both these seminars have already been published in English and are available to all who desire to acquire them. The third successive seminar had as its topic Environment and Communication. The fourth concerned itself with the topic Environment and Justice. The topic of this present fifth seminar is Environment and Poverty.

As is apparent from the preceding list of topics, the center of our concern focuses on the human person, who indeed lives within a specific terrestrial and natural environment. For, the environment receives its worth from—and receives its name in relation to—a distinct focal point, which it surrounds. This focal point is, in essence, humanity, or otherwise the human biotope—not unto itself, but as a human ecosystem, because humanity, of course, does not live nor is it able to live by itself in nature. Human beings live simultaneously and collaterally in this system, together with the multitudes of living plant and animal organisms, each of which thrives or, even better, survives in specific environmental circumstances. Humanity is dependent on a natural ecosystem in which the needs of all the living organisms that coexist are well balanced and served by each other. The disturbance of this balance within the ecosystem renders the survival of certain types of life difficult, while their possible extinction causes further disproportionate or asymmetrical growth and development of others. The result is the inability to meet their needs and, consequently, their death. The latter can often be seen as the desolation and the laying to waste of the ecosystem or as a significant change of habitat for other, usually more inferior, systems.

HUBRIS AND NEMESIS

In ancient Greek thought, immeasurable growth and excessive development is sometimes characterized as *hubris*—coinciding with elements of haughtiness and disrespect—while the consequence of this is brought on by *nemesis*. In Christian terminology, we speak of sin whose basic trunk, root, source, and point of departure is human pride, and whose wages are called, in one word, "death" (Rom. 6.23).

The topic of this year's seminar presents us with an inquiry about the meaning of the term "poverty." The fact that the term "poverty" is in contradistinction to the term "wealth," immediately guides our thoughts in the direction of ethics or its deontological character—namely, the study of our moral and ethical duty and obligation. For, in the purely biological sphere, where the selection of the good ethic and the bad ethic gradually occurs, we speak of sufficiency and insufficiency as an objective condition, independent of the willingness of the subject and, therefore, as a neutral ethic.

For example, we discuss whether in certain ecosystems nourishing or growing substances are either sufficient of insufficient, but we do not assign responsibilities to the living beings existent in them, which affect the sufficiency or insufficiency of the nourishing substances or any other living substance in general.

However, in any human society, it is stated that some of its members grow in wealth or otherwise live over and beyond sufficiency—that is to say, they possess more than they need. At the same time, others in society have less than the basic necessities that they require for survival, and we immediately question ourselves as to the reason for this unbalanced distribution of wealth. We begin thinking that perhaps the totality of available goods is sufficient for the totality of the needs. Consequently, perhaps the observable superabundance and surplus for some and lack of the same for others is the result of the blameworthy desires of the greedy, who abuse their abilities and strengths, because they are in possession—without real necessity—of the portion belonging to those who are lacking the basic necessities of life.

Consequently, the topic of poverty has many aspects and dimensions, and the appreciation and discernment of all these issues exceeds the boundaries of the topics of this present seminar. Besides, all of human history unfolds as a struggle to extend human power—both personal and collective—over the material goods of the earth, in a manner that would provoke the perplexity of every true philosopher—just as it did to primitive man living sufficiently in his natural environment—as to why this mad frenzy exists in humankind for the exclusive control of goods, which should be sufficiently available and shared by all.

REALITY AND IDEAL

For the duration of this present seminar, we shall confine ourselves to the investigation of the connection between the environment and poverty. One can study this specific topic from two points of view: the objective or the deontological. We characterize as objective that which we see in researching the sufficiency or the insufficiency of those goods that are naturally produced, and which by human activity are grown in order to satisfy human requirements. Let us set aside for now the views of economists concerning the boundlessness of human needs; these views do not speak about the natural, but about the psychological or spiritual needs of humankind. However, if we assume that, as the Church, we cannot sanction the boundless material needs and desires of the diseased human soul that are caused by greed, but can sanction only those that are genuine or natural needs (with a certain inherent alteration toward the common good), and if we compare the available means with those that are realistically existent—namely, those both naturally and ethically justifiable—then we will be able to confirm with much astonishment that the earth is capable of nourishing and generally satisfying all of these needs for the totality of its population.

The pessimistic theories of those who support opposing views have not been proven true. On the contrary, it has been proven that through technological and scientific progress in general, overabundant cycles of surplus crops have been achieved to the point that problems do not center on the lack of material earthly goods, but rather on their overabundant yield. Certainly, a balanced growth has not yet been achieved to a sufficient level causing certain regions to suffer because of overabundance and others because of deficiency and lack of goods. Nevertheless, this is not the result of the lack of human possibilities, but rather the lack of human will and desire to do what ought to be done, and the lack of human ethic and human organization and cooperation. In reality, this is an ethical or deontological problem, and not a natural one.

Perhaps there will be certain objections from among the pessimists, but we are of the opinion that these do not stand. Especially concerning the industrial and manufacture of goods, it is most apparent that the possibilities are great and can definitely satisfy all human material needs and wants. About natural goods and products, the information at hand speaks about a surplus of goods in technologically advanced countries. This fact indicates that through the use of modern methods and technology, sufficiency and surplus of goods that are available and manifest in all these counties can be readily increased and expanded to meet the needs of all peoples of the earth—if there exists the human will to do so. Consequently, this objective analysis leads us to the conclusion that this problem is the result of selfish and inappropriate (anti-deontological) human behavior, and not because of the lack of natural means.

MERCY AND CHARITY

We now come to the second point of view, the deontological or ethical view of this topic. The relevant positions and teachings of the Orthodox Church on these matters are, we believe, well known. The spiritual leadership of the Church, as well as every member who belongs to the Church, must sense that the needs of the least of the brothers and sisters of the Lord are their own needs, and they should work toward the satisfaction and fulfilling of these needs. The commandments of our Lord and the holy Apostles in regard to this issue are numerous. The beatitude "blessed are the merciful, for they shall obtain mercy" (Matt. 5.7); the beautiful parable of the last judgment, in which the primary criterion of judgment will be the loving behavior, or the lack thereof, of one human being toward one's fellow human being (Matt. 25.31); the words of the Apostle James, the brother of the Lord, who wrote that "pure and undefiled religion before God and the Father is this: to visit orphans and widows in their trouble, and to keep oneself unspotted from the world" (James 1.27); and likewise "if a brother or sister is naked and destitute of daily food . . . and you do not give them the things necessary for the body, then of what good is this?" (James 2.15–16). These are just a few examples of the positive Christian obligation to relieve the needs of one's fellow human being. Inversely, the explicit condemnation of greed as covetousness (Col. 3.5); the proclamation of the Lord in regards to the difficulty of the rich in entering the kingdom of heaven (Matt. 19.23); the marking of the desire of acquiring material goods and pleasures as the reason for wars (James 4.1–2), are but a few examples of the ethical unworthiness of the possessive ideology of the secular mindset.

This is how the Christian *phronema* finds itself close to the natural order of things, which, through the mouth of the Lord, is offered to humankind as the example of life: "Therefore do not worry, saying, 'What shall we eat?' or 'What shall we drink?' or 'What shall we wear?' For, the Gentiles seek after all these things" (Matt. 6.26–28). The Christians believe that God, who feeds them and clothes them, will give also to us the necessary things for our livelihood. This exhortation—an exhortation to trust in divine providence—that is to say, this faith in the love and the concern of God for us, does not revoke our obligation to work and produce that which is necessary. Rather, it also condemns the lack of faith and avarice, as well as the excessive and measureless occupation with this topic. Of course, the neptic Fathers—without despising material goods, but only through the grace of ascetic struggle learning to forsake the multitude of these and—preaching by the mouth of Abba Isaac the Syrian—declare that "it is clearly known that God and His Angels rejoice in caring for what is necessary, while the devil and his co-workers rejoice in resting."[2]

WANTING AND WASTING

This is how a Christian is guided toward a balanced use of the material goods of creation in temperance and contentment (1 Tim. 6.6–8), and journeys the royal middle road, praying with the author of Proverbs: "Give me neither poverty nor riches" (Prov. 30.8). Of course, humanity today "wishing to become rich," falls "into temptation and snares, and into many foolish and harmful lusts, which drown people in destruction and perdition. For the love of money is the root of all kinds of evil" (1 Tim. 6.9–10).

This excessive acquisitiveness in today's world is greatly responsible for a large part of the ecological destruction of our planet and, in the final analysis, proves to be at the expense of all humanity, including those who desire to enrich themselves. This is how, after a first phase of over-exploitation and luxurious living, the phase of deficiency and poverty necessarily follow suit, because in one period those goods that were to be used by the many were greedily used up by the few. This means that poverty is not the result of an objective insufficiency of resources, but a predatory exploitation, followed by the wasting and improper use of certain resources. Therefore, it is an ecological destruction on the one hand and an irresponsible and unequal division of goods on the other. The responsibility for both of these is borne by humankind, and we are obliged to help people to understand this in order to work actively toward

---

2. *Ascetic Treatises*, Homily 27.

the notion of the logical and reasonable use of the earth's resources, the non-disturbance of the ecological balance, and the preservation of the ability of our planet to yield and produce life, so that poverty may be abolished or at least diminished.

Summits, Events, and Ceremonies

Address at the Summit on Religions and Natural Conservation,
Atami, Japan, April 5, 1995

RELIGION AND CONSERVATION

The alliance of today's environmental issues for the protection of the environment with theological presuppositions may be likened to a paradoxical and even eccentric enterprise. In our perception, ecology, on the one hand, represents the pursuit of practical and desirable strategies. On the other hand, theology, or even theological cosmology, if used only as terms of expression, are understood by many people to be a kind of empirically interrelated, abstract, theoretical research. These disciplines, which refer to the association of dogmas and ideologies, have little or no regard for the practical aspect of life or religious issues.

Contemporary ecology, as subjective scientific research, often under the guise of crusading movements for the salvation of the earth's ecosystem, is an expression of the various activities that identify humanity's concern for practical application. Nevertheless, the rationale behind such a concern for environmental protection is often projected purely as logic of convenience. If we do not protect the natural environment, then our own survival will become increasingly miserable and problematic. Before long, the very presence of humanity on our planet will be threatened. Day by day, the degeneration and even the extinction of the human race become more markedly imminent.

COVENANT AND CONSERVATION

Within the context of this obvious logic, the natural environment is understood as an essential and acceptable party in a relationship—one might in fact speak of a "covenant"—with humanity for humanity's very survival. It is, however, a covenant or a context, which is comprehended only in terms of the environment's usefulness. On the human side, the problem is limited to how something is used. The origin or cause of physical reality is of no concern, nor is the search for a clarified "meaning" of cosmic propriety, harmony, wisdom or natural beauty. It is possible that the existential matter of nature was created by unknown "powers from above," possibly by inexplicable products of "chance," or even by an equally inexplicable automation, which exists intrinsically in the composition of matter. At all events, it is not the interpretation of cause and aim, which gives *meaning* to the existential matter, but rather, the meaning of existential matter for humanity is confined only to its *utility and advantage*.

Stemming from this logic of convenience, much of the ecological movement today demands that certain rules be set down for the manner in which humanity is to make use of nature.

Ecology aspires to be a practical ethic of human behavior toward the natural environment. However, as in any ethical system, ecology, too, raises the following question: Who determines the rules of human behavior and under which authority is this ethical system exercised? What logic makes these rules compulsory and what is the source of their validity? The correctness of an ecological ethic is made evident by its empirical usefulness. It is irrefutably logical that in order for humankind to survive on the earth, a covenant with the natural environment that subscribes to human survival must be negotiated.

However, such a rationale, which is founded on the intentional utilization of what is convenient and advantageous, has led in the first place to the destruction of the natural environment. Man does not destroy the environment because he is motivated by an irrational self-gratification; rather, man destroys the environment by trying to take advantage of nature in order to secure more conveniences and comforts in his daily life. The logic of the destruction of the environment remains precisely the same as that of the protection of the environment. Both "logics" confront nature as an exclusively utilitarian commodity. They give it no other meaning. They are motivated at the same level by an ontological interpretation of physical reality—which is, in fact, more correct—toward the void, which comes out of a desire to deviate from every ontological interpretation.

In this way, the difference that exists between the two "logics"—namely, between that of the destruction and that of the protection of the ecosystem—is ultimately only quantitative. Ecologists demand limited and controlled exploitation of the natural environment—namely, a quantitative reduction—that would also permit its further, longer-termed exploitation. They ask for the rational limitation of non-rational usage, in other words, a kind of consumerist rationalism, which is "more ecologically correct" than the consumerist rationalism of today's exploitation of nature. They are, in the final analysis, asking for consumerist "temperance" of consumerism.

Yet, who will determine absolute quantitative "ecological correctness" within the context of a single monolithic logic of convenience? Moreover, with what means will this occur? This demand, while appearing to be extremely rational, is, by definition and in practice, irrational. By definition, it is self-contradictory because consumerism cannot come into conflict with itself. In addition, it is irrational in practice because the majority of the world's population does not normally and willingly deprive itself of conveniences and comforts. In reality, only a small minority of "civilized" societies has secured its conveniences and comforts through the destruction of the environment.

THEO-LOGICAL LOGIC

In order that the demands of ecologists be seen as viable, a new logic is required, one that is able to replace the logic of convenience. The demands should be founded on entirely different principles. One example of this might be in the altruism of someone who cares about the destiny of future generations; another might be in the false quest for some "quality" of life that is not judged by consumerist ease or abundance theories. In other words, the demands must be for universally accepted goals whose basis rejects the utilization only of what is convenient. Moreover, in defining these goals, it appears from all accounts to be impossible for everyone to agree uniformly on rational criteria. People, therefore, must come up not only with different demands, but also with an entirely new hierarchy of demands. In addition, different demands arise only when in the consciousness of the people, the world and nature acquire another meaning that is not exclusively oriented toward convenience.

Christianity, along with other monotheistic religious traditions, preserves the attitude that the physical reality is not exclusively a convenience. According to these monotheistic traditions, the world is a creation of God. The use of the world by human beings constitutes a practical relationship between humanity and God because God gives and humanity receives the products of nature as an offering of divine love for the sake of the world. My friends of the Islamic faith tell me that in the Qur'an it is said that all animals live in community and that they are known by and accountable before God.[3] The Muslim faith also denounces the arrogance of those who treat the rest of creation without respect: "Do you not see that it is God whose praises are celebrated by all beings in heaven and on earth, even by the birds in their flocks? Each creature knows its prayer and psalm, and so too does God know what they are doing. And yet, you understand not how they declare His glory."[4]

HUMANITY, COVENANT, AND ECOLOGY

Within this picture of all creation arising from a loving Creator God, the question has to be asked: "What of humanity?" What does our faith ascribe to human beings? The answer is clear and the consequences immense. The eighth Psalm addresses this question directly:

> When I look up at your heavens, the work of your fingers,
> at the moon and the stars that you have established;
> Ah, what are human beings that you should
> spare a thought for them,
> Mortal beings that you should care for them?
> Yet you have made them a little lower than God.
> You have crowned them with glory and honor,
> You have given them dominion
> over the works of your hands,
> And set all things under their feet,
> sheep and oxen, all these, yes, and the wild animals too,
> the birds of the air, and the fish of the sea,
> whatever travel the paths of the ocean.
>
> (Ps. 8.3–8)

---

3. Sura 6.38.
4. Sura 24.41.

We have already mentioned how important it is for an agreement or covenant to exist between humanity and creation for the eternal survival of both. This notion of covenant stems from a figure of fundamental importance not only to Christianity, but also to Judaism and Islam. We are referring to the patriarch Abraham. God established with Abraham a holy covenant that confirmed a permanent relationship between him and his chosen people. Jewish scholars say that Judaism sees Abraham as the founder not only of its faith but also of its people. They consider themselves as "the people of the Covenant." In the book of Genesis, the Lord says to Abraham, "Leave your country, your family and your father's house, for the land that I will show you. I will make you a great nation; I will bless you and make your name so famous that it will be used as a blessing" (Gen. 12.1–2).

Christianity sees Abraham's covenant as a forerunner of the New Covenant, established by Jesus Christ, in which Christian believers become spiritual descendants of Abraham by divine adoption. This is clearly revealed in the hymn known as the *Magnificat*, cited by the Evangelist St. Luke. It was sung by the Virgin Mary as a song of praise for the gift of Jesus. In this hymn, Mary describes how God will bring down the mighty from their seats and exalt the humble and meek. He will fill the hungry with good things and send the rich empty away. At the end of her song of praise, she cries: "He has come to the help of Israel His servant, mindful of His mercy, according to the promise that He made to our ancestors, of His mercy to Abraham and to his descendants for ever" (Lk 1.54–55).

My Muslim friends say that Islam also sees Abraham as the founder of the Kabba at Mecca, and that the Qur'an holds him to be the first true believer and a model for all believers. Central to that Covenant is the belief in the one God-Creator and sustainer of all that has been, is, and will be. Nothing exists but for the will of God. As the Book of Genesis says: "In the beginning, God created the heavens and the earth. Now the earth was a formless void, there was darkness over the deep, and God's spirit hovered over the water" (Gen. 1.1–2). This reality is echoed over and over again in the Qur'an.

Christianity inherited this Old Testament tradition through the teachings of Jesus, who compared the value of human beings to sparrows, both loved and cared for by God. Yet human beings are a hundred times more important (Matt. 6.26). It is also the case that the sense of human beings being part of a bigger picture, a greater purpose of God, is to be found equally emphasized within both Jewish and Christian texts. For example, in the Torah, in Genesis 9, the Covenant with Noah after the flood is not just made with Noah and his descendants—namely, with the human race alone—but with all life on earth. Similarly, when St. Paul speaks in Romans 8 and in Colossians 1 about the purpose of the life, death and resurrection of Christ, he does not regard these as affecting only human beings, but as occurring for the sake of all life on earth.

COMPASSION AND CONSERVATION

Ultimately, for Christianity—and, as we understand it, for Judaism and Islam as well—humanity is the most important or most significant species. Nevertheless, with this reality come particular responsibilities. We are told that in Islam this is expressed by the notion of humans being *khalifas*. In this tradition, a *khalifa* is the vice-regent—someone appointed by the Supreme Ruler to have responsibility over a given area in an empire. We have been told that the Qur'an uses this phrase as a description of the role of humanity. God has given humanity this authority. However, it is only to be used on God's behalf, not for our own ends and ambitions. Any abuse of this power, any wanton or wasteful use of the world's natural resources, is repugnant to God.

The belief that God's love underpins the natural world is borne out in the teaching that God Himself is also bound by His love for the world. Therefore, when Abraham stood before God after God declared that He was about to destroy the sinful city of Sodom, Abraham demanded of God that He give justification for this act, saying, "Shall not the judge of all the earth do justice?" (Gen. 18.25). Christianity, along with Judaism and Islam, also acknowledges that humanity also has been given dominion over nature but always within the context of love, justice, and compassion. We live in a tension between the scale of our power and the limits imposed by love and conscience.

In the Christian faith, we are taught that Christ is part of the creating Trinity of Father, Son, and Holy Spirit. Thus, with St. Paul, we can describe Christ as Creator who was emptied of His power, and came to earth as a child—as a weak and defenseless child. The Word of God took upon Himself the role of a speechless servant. So at the heart of Christianity is a Creator who becomes a creature; the mighty Lord is born as a child; the Master of all now becomes the servant of all.

One of the most challenging aspects of the teaching of Christianity, and perhaps also of Judaism and Islam, is the belief that we are able to choose to disobey God. Our freedom under God is immense. Our capacity for good is vast. Our capacity for evil is likewise vast. And often this evil comes about because of foolishness, greed, pride, or ignorance. Yet the end result is always the same, namely destruction.

COMMUNICATION AND CONSERVATION

Two fundamental consequences emerge from the understanding of our traditions. The first consequence is that the world is not meant to be used by humans for their own purpose, but it is the means whereby humans come into relationship with God. If humans change this use into egocentric, greedy exploitation, into an oppression and destruction of nature, then humanity's own vital relationship with God is denied and refuted, a relationship predestined to continue into eternity.

The second consequence is as follows. The world as a creation of God ceases to be a neutral object for our use. The world incarnates the word of the Creator, just as every work of art incarnates the word of the artist. The objects of physical natural reality bear the seal of the wisdom and love of their Creator. They are words of God calling humanity to come into dialogue with divinity.

It is, therefore, a fact that humankind today must change its position with regard to the natural environment. This is a necessary prerequisite for humankind changing the meaning that it gives to matter and the world. Ecology cannot inspire respect for nature if it does not express a different cosmology from that which prevails in our culture today, and one that is liberated equally from naive materialism and naive realism.

Only when human beings confront matter and nature as the work of a personal Creator does the use of matter and nature in turn establish a true *relationship* and not a monolithic domination of man over physical reality. Only then will it be possible to speak about an "ecological ethic," which does not borrow its regulatory character from conventional, rational rules, but out of the need for a person to love and be loved within the context of a personal relationship. The reason there is beauty in creation, therefore, is one of amorous love; it is an invitation from God to humanity; it is a call into a personal relationship and to communion of life with Him; it is a relation that is vital and life giving. Contemporary ecology could therefore be seen as the practical response of human beings to God's call, the practical participation in our relationship with Him.

Is the ontological clarification of the *meaning* of matter and of the world sufficient to prompt a different relationship between contemporary humanity and matter and the world? Of course not. In order for humanity to confront nature with the respect and awe with which we confront a personal artistic creation, the above-mentioned theoretical clarification must become a powerful source of knowledge and a society's position. At this point, the role of social dynamics, as referred to in ecclesiastical tradition and community, may be determined as long as the ecclesiastical conscience is purged of its liberation to an ideological structure and from its inert retirement in the restoration of institutional formalisms.

There is a tension that exists between the ability to do the will of God and the ability to rebel. Moreover, alongside this tension lies the ability to return to the right path, to follow the again the path of God revealed in the Scriptures. This is why we can look upon the present state of the world with comprehension and distress, but also with a sense of hope: *comprehension*, because we can see what happens when people forget that this world is not ours but God's; *distress* when we see the consequences of injustice and wrongdoing, of pride, foolishness, and greed, which go against the teachings of God; and most of all with *hope*, for we all believe in faith and confidence that if we turn again to God, God will meet us and bring us home to Himself. By returning to God, we may start again. The past can be forgiven and we can hope—truly and sincerely hope—that things can be fundamentally different.

If there is a future for the ecological demands of our contemporary times, this future is based, we believe, in the free encounter of the historical experience of the living God with the empirical confirmation of His active word in nature.

Address to National Religious Leaders, Hong Kong,
November 6, 1996

CONSERVATION AND COMPETITION

When closed within a very narrow environment, where they know only those of their own race and religion, people can create within themselves imaginary and deceptive images of others who belong to other races and religions. However, Odysseus of ancient times, "saw cities of many people and acknowledged their intelligence." He was able to ascertain, as we the modern day examples of Odysseus are able to ascertain with him, that all of us descend "from one human race," as the Apostle Paul wrote, and consequently, that we are all brothers and sisters.

This brotherhood unites us and at certain times separates us. This happens when we heartlessly treat our siblings unfairly when it comes to dividing up a rich inheritance—namely, this small earth which hosts us and which perpetually moves within the unfathomable universe—given to us by the Father, our good God, who is ever-beneficent and possesses superabundance. For our loving Father has organized everything so well in order that this Garden of Eden, our earth, might sufficiently provide for all the needs of those who live upon it. A simple demonstration of this would be the voluntary destruction of surplus production of goods from the developed nations, while for others there is a scarcity of goods for consumption, even to a dangerous level. A further example of this would be the total waste in inconceivable proportions of work and money for the production of weapons systems and their catastrophic deployment. Clearly, if all of these efforts were put toward the good of humanity, life on our planet would be much better.

Here in the peaceful and progress-oriented city of Hong Kong, you live within an efficiency and creativity of human effort, and you well understand the truth of our words. They are the words of a father or brother who beholds with sorrow some of his children destroying one other. He pleads with the rest of his children not to be misguided toward such situations, but rather to resolve their differences peacefully through mutual withdrawal. There is always a solution that is more advantageous than the one that leads to conflict. The obstacle for discovering this solution is not found in our external world, but rather within ourselves, because all our decisions spring from deep inside us.

We bless your peaceful place. We greet you with love and emotion at this gracious welcome, which you have reserved for our Modesty and our entourage. We wish you health, peace, and progress. May you live and be a universal example of a peaceful and civilized community, an example of the effectiveness of freedom and of noble competition. For in this blessed place of yours, virtually all of the peoples of the earth meet and work together, coexisting peacefully in spite of their various differences. This is because, among yourselves, you have accepted your neighbor and his or her right to be as he or she is. It would be an oversight if the effectiveness of this attitude, which is characterized as freedom, mutual respect, and labor were not duly noted.

Address to the Bankers Association, Athens, May 24, 1999

ENVIRONMENT, PEACE, AND ECONOMY

We are truly overjoyed with the present gathering. We are assembled here in love, beyond our regular professional obligations, and this reality adds a supreme sensitivity to the atmosphere that is created among us, something that already constitutes a spiritual environment. For, we are not standing opposite one another in order to exchange some professional agreement or contract, but rather we are standing together with one another as coworkers in a common good.

The first part of this common good is getting to know one another and establishing an inviolable bond of friendship, which marks the presupposition of any harmonious cooperation. The second part of this common good is the study of our mutual interests in order to examine how we might successfully achieve our goals. We are deeply grateful for your invitation to this gathering in order to contribute whatever we can to your goals. We are also grateful for your evident love, which reveals feeling hearts and thinking minds.

SAVING MONEY AND SERVING PEOPLE

If one considers matters superficially, then one would say that there are no common points that we share. For, you work in the sphere of matter and money, while we work in the sphere of the spirit. Nevertheless, a deeper study of matters persuades us that things are not quite this way. Indeed, we are bound by a fundamental common element, namely our interest in the human person. Let us develop what we mean by this.

The literal sense of money in the Greek language implies its inner meaning, namely something that can be used. However, wherever we speak of usage, we are implying someone who uses as well as a goal for this usage. This signifies that human beings do the using, while their aim is to cover material—and more rarely, when these are related to material goods—to meet spiritual needs. Therefore, money is a tool in the hands of human beings, and its purpose is to serve human needs. This means that, inasmuch as you are occupied with money and preoccupied in dealing with money, in the final analysis you are—or should be—serving humanity. For, money is surely not an end in itself. Nor is money the object of some lifeless treasuring up. Rather, money is a catalyst that facilitates exchanges; it is an ever-moving catalyst, which, when properly used, offers a sense of satisfaction on the persons through whose hands it is exchanged. This happens even in the case of someone who painfully counts it out; for, it is also a matter of reward for that person. Consequently, in meeting material human needs, you are ultimately serving human beings.

From a different perspective, and with quite different means, we, too, serve human beings, seeking to meet their spiritual needs. Nevertheless, we know that the human person is a psychosomatic being, comprising spirit and matter, and that his or her physical needs must be met in order for that person to stay alive and enjoy spiritual needs. Therefore, if a person does not satisfy one's material needs, whether individually or through other people such as yourselves, then we are required to assume responsibility for their needs voluntarily and charitably. The Gospel command is very clear in this regard: Give to those who do not have; take care of the orphans and the widows; feed the hungry; heal the sick; help the helpless. The Church does not ignore material needs, but it incorporates these within an appropriate hierarchy, wherever necessary giving priority to the primary and ranking lower the secondary, in order to meet the basic needs of those who are lacking in material and spiritual things.

Bearing this in mind, when we notice our fellow human being lacking even that which is given freely by God to all, namely life-giving oxygen and clean air, simply because other human beings consume the available oxygen and pollute the atmosphere with harmful waste, we come to the conclusion that the abuse of our technical resources is morally impermissible. This is because it results in the deprivation of life-giving natural goods among some people and in the enrichment of others through financial profit. So, then, just as it took many years to establish free shipping trade, namely to accept that the seas are open for the use of all humanity, in the same way efforts are required to establish people's right to breathe clean air and to inflict proper penalties on those who pollute the air. Because clean air and the preservation of the natural environment in general are necessary for the healthy existence of each person, these are also the obligation of each person.

Unfortunately, people have not yet become conscious of this obligation, and so we have mobilized ourselves out of the above-mentioned sense of charity to assume responsibility to enlighten everyone—for, we all consume air and water—concerning the harmful consequences of our pollution for humanity. For such attitudes and actions that result in pollution, also conceal an ethical insensitivity. Our efforts aim at sensitizing people's conscience and at voluntarily curtailing mass actions of pollution for the sake of social obligation. Indeed, the arousal of people's conscience constitutes a prerequisite for us to stop actions that destroy the natural environment.

Therefore, we propose that we might spend a little more time on this critical issue. The aim is not to idolize the environment but to serve humanity. In particular, we would like to develop the aspect of the environment that relates to peace and war. For, if during times of peace, when the protection of human beings is perceived positively, ecological issues are acute, then how much more critical are these issues during times of war, when the extermination of others and the destruction of their environment are the unfortunate objective?

If we study in detail the conditions of war described in the epics of Homer,[5] and then proceed to compare these to contemporary situations, we will be surprised at the insignificant impact of war at that time on the environment, at least by comparison with the tragic effects that we witness today. Indeed, if we consider the consequences of war at different historical periods, then we shall also observe the sad reality that, the closer one

---

5. The Iliad and the Odyssey date from around the ninth or eighth centuries BC.

comes to our period, the more dramatic the effects of military clashes on the natural environment.

AN ENVIRONMENTAL DECALOGUE

An enumeration of the precise impacts and effects of war on the environment is not easy for someone who is not a specialist or scholar in the field. The following is a rough outline of the most obvious effects.

A significant number of fatalities, leading to the disruption of families and sometimes even societal structures in communities.

A vast number of casualties, with the same consequences and additional expenses for their health care and preservation, which are costs subtracted from other areas of life.

An unknown number of those who succumb to illnesses as a result of military pollution of the environment by means of chemical gases, radioactive substances, fire and decomposition.

An indeterminable number of spiritual wounds resulting from the cruelty of war in numerous communities, which thereafter foster anti-social feelings and disturb the human environment.

Long-term pollution of the region by the wastes that result from military machinery. For instance, huge amounts of air pollution are dumped as a result of the thousands of hours in military flights over a given region, the shooting of jet-propelled rockets, the sailing of navy ships, the movement of army and administrative vehicles. All these pollutants travel not only on the specific region of the war but throughout the neighboring regions, even reaching distant territories where they affect good and evil people alike.

An especially affected and gravely polluted region where the conflict occurs, as well as the nearby regions, which suffer from the by-products of explosives, charged and diffused, in the form of radioactive shells of ammunition, rockets, bombs, and other modern weapons of mass destruction. This pollution is conveyed throughout the region not simply by means of the air, but also by means of water, thereby affecting areas that are not even involved in the conflict.

Equally tragic is the destruction and discharge that occurs in tanks and reservoirs, as well as in factories of chemical products. If one considers the detailed, strict, and careful measures in place for the safe transferal, storage, and development of these dangerous products until they are transformed into inactive and harmless products for general consumption, and if one considers how all of these products are exploded into the air together as a result of military bombing, then it is impossible not to be overwhelmed by a sense of sorrow. It should be noted that the involuntary recipient of these pollutants is not only the military opponent; indeed, it is not only the non-combatant civilians among the enemy, which in any case, according to international regulations of war should not be the target of military attacks. Rather, the recipients also include those populations beyond the borders of the country at war, whether in neighboring or more distant regions. Indeed, it even includes the soldiers themselves who are causing the destruction! The unforeseen effects of the heavy pollution caused by war may appear on the other side of the planet, including the country of those instigating the military attack! For, the interconnections and mutual influences within nature are vast and often inscrutable. It is sufficient to recall the example of scientists, that the fluttering of a butterfly's wings in Japan is sufficient to be the cause of rain in America.

Still more tragic is the radical and often irreparable destruction wrought upon local human and other ecosystems, which suffer as a result of the effects of war. Ecosystems in the oceans, in rivers, and in lakes are killed by explosions of bombs and mines; terrestrial ecosystems are destroyed and annihilated not only by explosives but also by fires, which further clear forests and level homes; road systems are dismantled and organized human lives regress to the conditions of the past.

In addition to all this, cultural monuments are destroyed or damaged, so that civilization itself suffers a lasting blow, as the organization *Europa Nostra* has declared through its resolution dated April 29, 1999.

Finally, the spiritual atmosphere is also inundated by boundless falsehoods of propaganda; passions are cultivated in people's souls; hatred and an attitude of destruction are justified; the effects of this spiritual "pollution" are manifested at any point in the world, irrespective of distance, for example in a school where a young child learns to develop racist attitudes against invisible enemies or even among visible schoolmates. This psychological pollution, which adversely affects the human environment, is especially important for us, although usually overlooked by those who deal with environmental issues.

THE IRRATIONALITY OF WAR

This general decalogue of environmental effects that result from contemporary war indicates the irrationality of military conflict, which can only be explained as a paranoid act. For, while war is instigated in order supposedly to protect certain people who are provoked by their unjust treatment by other people, nevertheless the unjust treatment is extended to include numerous other people. Moreover, while the injustice against which people seek to protect themselves is connected to some financial or territorial gain, nevertheless vast amounts are expended for the destruction of the enemy and only minimal amounts are left for the consolation of the aggrieved. Perhaps an inverse apportionment and expenditure might have successfully resolved many conflicts.

Therefore, the irrationality of war is evident from its effect on humanity and on the natural environment. It is our duty to intervene, wherever possible, to persuade those who are responsible for making decisions in order that they might seek peaceful resolutions to human problems. With good will and the proper effort, such solutions can be found. The choice of military violence as the sole method for resolving or imposing issues betrays a lack of satisfactory imagination and reveals intellectual laziness as well as confidence in the erroneous notion that evil can be corrected by evil.

As heralds of the Gospel truth, which is the only complete truth, we repeat the words of the Apostle: "Do not be overcome by evil, but overcome evil with good" (Rom. 12.21). We conclude with this exhortation, adding only our fervent prayers that irrational wars may cease as soon as possible and that the almighty and beneficent Lord may grant everyone the wisdom to understand that war is an impasse. May the same Lord decrease as much as possible the dark consequences of military attacks and grant peace to all peoples.

Remarks to the Board of the Sophie Foundation, Oslo, June 12, 2002

HUMAN ENVIRONMENT AND NATURAL ENVIRONMENT

Interpersonal relationships are among the more significant capacities of human nature. Indeed, we could say that human communication—especially with God and one's fellow human beings—constitutes the primary element of human existence. For, it is not possible for us to conceive of humanity without personal relationships; just as God is inconceivable as an impersonal Being, but only as Trinity.

A human being exists as a human person, namely in relation to other human beings. Therefore, apart from the natural environment, we should also speak of a human environment. The latter is no less significant than the former. The present banquet reminds us of the warmth and joy of a human environment. Unfortunately, in contemporary societies, this human environment is no less problematic than the natural environment.

It is our duty to improve both environments, introducing elements of love, understanding, reconciliation, tolerance, communion, dialogue, and respect for one another, in order to render human life more personal.

Message to the Caretakers International Youth Summit, Oregon, June 2, 2005

YOUNG PEOPLE CARING FOR THE EARTH

It is with particular joy that we welcome the opportunity to offer this brief message of paternal greetings and fervent wishes to the committees and participants of the 2005 Youth Environmental Summit of Caretakers of the Environment International.[6] For some time now, we have followed with great interest the unique initiatives of Caretakers International and recognize the significance—in fact, we could say, the sacredness—of their educational projects with young men and women throughout the world in an effort to preserve the natural creation and promote greater cooperation among the various sectors of the community for the protection of our planet.

This year, your summit will be held in the beautiful surroundings of Willamette University within the region of Salem, Oregon. The theme of your deliberations and challenges has been wisely defined: Forging New Partnerships with the Economic Community.

It is critically important for all of us to realize—and to do so from as tender an age as possible—that only by fostering open dialogue and sincere collaboration will we also be able to awaken people's conscience with regard to what we are doing to God's world. Whether your teams will discuss such issues as the reduction of solid waste, the practice of recycling, the improvement of water quality, the conservation and development

---

6. A global network of secondary school teachers and students active in environmental education.

of alternative forms of energy, or the preservation of endangered species, we know today that we cannot be successful unless we work together with one another and unless we involve businesses and the economic community.

"ECOLOGY" AND "ECONOMY"

The terms "ecology" and "economy" share the same etymological root. Their common prefix "eco" derives from the Greek word *oikos*, which signifies "home" or "dwelling." How unfortunate and indeed how selfish it is, however, that we have restricted the application of this word to ourselves. This world is indeed our home. Yet it is also the home of everyone, as it is the home of every animal creature, as well as of every form of life created by God. It is a sign of arrogance to presume that we human beings alone inhabit this world. Indeed, it is also a sign of arrogance to imagine that only the present generation inhabits this earth.

Ecology, then, is the *logos* or study of this world as our home, while economy is the *nomos* or regulation, namely the stewardship of our world as our home. How we understand creation will also determine how we treat the natural environment. Will we continue to use it in inappropriate and unsustainable ways? Or will we treat it as our home? And as the home of all living creatures? Will we, to adopt the words of the Psalmist, remember that "everything that breathes praises God" (Psalm 148)?

We believe in you—in the younger generation that has learned how to respect God's creation for itself, for the protection of the environment, for the enjoyment of future generations, and for the glory of God. May the grace of God inspire and guide you as you work in your teams in order to meet the challenges of your projects during this Summit.

Message to the International Conference on Peace, Thebes, Greece, October 13, 2005

ENVIRONMENT AND PEACE

It is indeed a particular pleasure to greet the organizers and participants of the international conference on Environment and Peace, organized on the occasion of the 60th anniversary of the founding of the United Nations, as well as in connection with the inception of two decades proposed by the United Nations on education regarding sustainable development and the program called "Water and Life." This is indeed a historical gathering of esteemed religious leaders, governmental authorities, and academic scholars. The fact that the venue is the historical monastery of St. Lukas, within the sacred Metropolis of Thebes and Levadeia, demands of us an attitude of prayer as we reflect on God's creation.

Over the last decade, it has been a privilege of our Ecumenical Patriarchate to initiate sea-borne symposia—five to date—on the themes relating to the preservation of rivers and seas, including the Mediterranean Sea, the Black Sea, the Danube River, the Adriatic Sea, and the Baltic Sea, organized by the Religious and Scientific Committee under the inspired leadership of His Eminence Metropolitan John of Pergamon. We have learned that it is critical that our efforts to protect the natural environment be interdisciplinary. No single discipline or group can assume full responsibility for either the damage wrought on created nature or the vision of a sustainable future. Theologians and scientists must collaborate with economists and politicians if the desired results are to be effective.

Moreover, we have learned that environmental action cannot be separated from human relations—whether in the form of international politics, human rights, or peace. The way we respond to the natural environment directly reflects the way we treat human beings. The willingness to exploit the environment is revealed in the willingness to permit avoidable human suffering. All of our ecological activity is ultimately measured by its effect on people, especially the poor. Extending our concern and care to nature implies and involves changing our attitudes and practices toward human beings. The entire world is a gift from God, offered to us for the purpose of sharing. It does not exist for us to appropriate, but rather for us to preserve.

COMMITMENT TO CHANGE

In our efforts for the preservation of the natural environment, how prepared are we to sacrifice some of our greedy lifestyles? When will we learn to say: "Enough!"? Will we direct our focus away from what we want to what the world needs? Do we endeavor to leave as light a footprint on this planet for the sake of future generations? There are no excuses for our lack of involvement. We have detailed information; the alarming statistics are available. We must choose to care. Otherwise, we are betraying our human rights. Otherwise, we are aggressors.

"Blessed are the peacemakers; for they shall be called sons of God" (Matt. 5.9). To be sons of God is to be fully committed to the will of God. It implies moving away from we want to what God wants. It means to be faithful to God's purpose and intent for creation, in spite of the social pressures that may contradict peace and justice. In order to be "peacemakers" and "sons of God," we must move away from what serves our own interests to what respects the rights of others. We must recognize that all human beings, and not only the few, deserve to share the resources of this world.

"Making peace" is certainly painstaking and slow work. Yet it is our only hope for the restoration of a broken world. By working to remove obstacles for peace, by working to heal human suffering, by working to preserve the natural environment, we can be assured that God is with us (Emmanuel), that we are never alone, that we shall inherit both this world and the kingdom of heaven. For then, we shall be worthy to hear the words of Christ: "Come, you who are blessed by my Father. Inherit the kingdom that was prepared for you from the creation of the world" (Matt. 25.34).

7

"All in the Same Boat"

Intergenerational and Interdisciplinary Solidarity

SUMMER SEMINARS AND RSE SYMPOSIA

Remarks on Official Visit of HRH Prince Philip,
Duke of Edinburgh, Istanbul, Turkey, May 31, 1992

ICONS AND RELICS

God created humanity to serve as a king of creation, not for any individualistic exploitation of it that results in destruction, but for the enjoyment of a peaceful and fruitful life in it in harmony with the other creatures, plants, or animals. Nevertheless, we have recently experienced a dangerous development, arising from our senseless and often selfish use of natural recourses. The environment, as it is presented to us today, appears to resemble the "beast" in the Book of Revelation (Rev. 12.4), which waits to devour the newly born child of the woman. The only difference is that the beast of the environment, in a metaphorical sense, expects to devour everyone, and not just the newly born.

If we are to avoid this pressing danger, then we urgently need to restore the proper relation between humanity and the environment. This reconciliation has always been facilitated by the favorable position of nature itself toward humanity on the one hand, and the means that humanity has at its disposal on the other.

We often speak about respect for the human being as the icon of God. And this is correct. Yet we should not separate this respect from the respect that is due to the whole of the physical environment because it is obvious that the environment and its inhabitants are in constant and mutual interaction, as we pointed out in our Christmas message.

Allow us, Your Royal Highness, to refer to an example of a contemporary ascetic on the Holy Mountain, who made the following poignant comment: "We venerate the clothing of St. Nektarios,[1] because the saint used to wear it. Is it not much more fitting, that we should also venerate the flowers and the plants? After all, they enshrine within themselves the energy of God." It is because the true monastic has the measure of life that a monk or a nun will never turn either to idolatry or to pantheism. The monastic respects the whole of creation without attributing worship to it. Worship belongs to God alone.

Closing Ceremony, Symposium II, Thessalonika, Greece,
September 28, 1997

RE-ENVISIONING ECOLOGICAL HISTORY

The first page in the history of saving the Black Sea from ecological devastation has been written. Intentions have been sensitized; however, appropriate measures must now be taken. The baton has been passed, and each of us returns to his or her own main occupation. This does not mean an abandonment of the problem. It means watchfulness with preparedness for every assistance.

The account of the transactions and their conclusions will not occupy our Modesty at this present festal closing session. Instead, we wish to turn our attention within, to ourselves. For, as human beings, we are both the reason for the various ecological problems and the receiver of the results from these. Yet, it ought not to escape our attention that we are ourselves an "environment" for our fellow beings. Each of us, as a dweller in the wider ecosystem, is in this respect an environment for our fellow human beings. Human, it is true, but an environment. As a result, it is not sufficient that we secure the best terms for the natural ecosystem, which surrounds humankind, and with which we usually concern ourselves. Instead, it is imperative that we secure that human behavior, which is the best for human symbiosis as well. Unfortunately, the ecological problems relating to the natural environment are usually provoked by human

---

1. St. Nektarios of Aegina (1846–1920) is a popular and miracle-working saint.

actions, but despite the fact that this is serious, it is frequently of less importance in comparison with all the ills that humanity directly provokes against its fellow human beings.

REWRITING THE ECOLOGICAL STORY

Ecological adulteration is an imminent danger, a danger threatening both nature and humankind. Unfortunately, however, it is human beings who by means of other human beings are frequently not simply a danger threatening them, but the cause of a catastrophic event. History overflows and our days are equally full of persecution, oppression, genocide, execution, destruction and plundering of man by man. Thus, it is not just the damage to nature, resulting from greed and indirectly harming the members of human society, that is a serious problem. It is also the direct harm brought upon human society, resulting from various inhumane motives and developing within diseased and fanatical souls, which should be the focus of our attention. Consequently, we ought to turn our interest to these people as forming the environment of the rest of our fellow human beings and as creating for them good or, occasionally, unbearable conditions of life, and to investigate in what ways we are able to improve the conditions of man's dwelling alongside one's fellow man.

With full awareness, we have sidestepped the issue. However, if we do not change within ourselves the attitude of our heart toward our fellow human beings from an attitude of indifference or even enmity to an attitude of friendship, cooperation, and acceptance, then we will achieve nothing in the confrontation of the ecological problems of worldwide interest.

Having the hope that the heavy clouds of conflict among people will be dispelled, and daily offering fervent prayers for this cause, with gratitude and paternal love we bid farewell to the beloved participants of this symposium. And we entreat the Lord to protect and shield them all during their return journey and throughout all the days of their life that lies before them.

Foreword to Proceedings of the Fifth Summer Seminar, *June 2000*

FRAGMENTATION AND ISOLATION

The deterioration and, in certain parts of the planet, destruction of the environment constitutes an undoubted reality for our contemporary world. The Ecumenical Patriarchate follows with grave concern this deterioration of the ecological crisis, as well as the development of a tendency of an isolated and exclusive response on the part of various scientific branches and specific efforts to control this problem. As we have repeatedly underlined, the fragmentary examination of this problem succeeds only in a partial consideration and response.

In order, therefore, to examine and encounter comprehensively the environmental crisis that we are facing, in relation especially to the problems of poverty and social injustice, the fifth annual Summer Ecological Seminar was held June 14–20, 1998, in the Holy Patriarchal and Stavropegic Monastery at Halki on the theme "Environment and Poverty: legal dimensions and moral responsibility." Representatives of Churches and other religions, environmental and governmental authorities, as well as scientists and scholars, presented their views, participated in captivating discussions, and produced common documents, which comprise valuable material for a comprehensive approach to the issue at hand.

POVERTY AND ENVIRONMENT

For, as one of the more serious ethical, social, and political problems, poverty is directly connected to the ecological crisis. A poor farmer in Asia, in Africa, or in North America will daily face the reality of poverty. For these persons, plastic is not harmful to the environment and the destruction of the forests, but rather to the very survival of themselves and their families. Terminology such "ecology," "deforestation," or "overfishing" is entirely unknown. The "developed" world cannot demand from the developing poor to protect the few earthly paradises that remain, especially in light of the fact that less than 10 percent of the world's population consumes over 90 percent of the earth's natural resources.

The present situation reminds us of the poor widow in the Gospel, who made her small offering in the treasury; yet, this contribution was the equivalent of her entire possessions: "For, all of them have contributed out of their abundance; but she out of her poverty has put in everything that she had, all that she had to live on" (Mark 12.44). Therefore, we are not justified in demanding the poorer nations to offer huge sacrifices, when in any case these contribute far less than the developed nations to the pollution and crisis of the environment. Instead, we ought to assume our own responsibilities and contribute to the solution of the environmental crisis in accordance with our possibilities as financially stronger nations in order to wipe out poverty as well.

Remarks at the Academy of Sciences, Symposium III, Vienna,
October 19, 1999

SCIENCE, SCHOLARSHIP, AND STATE

Expressing deep gratitude for the invitation to address you from this distinguished podium, allow us first of all to express my warmest greetings to all of you and especially to His Excellency the Minister of the Environment of the friendly country of Austria, in whose competency lies the subject of this third international symposium, Religion, Science, and the Environment, and this year's sub-theme "Danube—River of Life." As we all know, "the teaching of the wise is a fountain of life . . . and to know this law is a privilege of a good mind" (Prov. 13.14–15).

Therefore, this temple, too, in which the best minds of your dear and hospitable country are sheltered, constitutes a source of life, the starting point for Laws inspired by the wisdom concentrated here. Because decrees passed by parliaments are not the only laws governing people's minds; there are also the ideas and arrangements proposed by wise men and women. It could be said that the citizen is more easily convinced to comply with what the wisdom of the scientist and scholar recommends, than to obey the commands of the constitutional legislator. This is why this symposium, although possessing no state authority whatsoever, is seeking to convince the citizens to cooperate in the effort to keep the Danube clean and to maintain a balanced and undisrupted environment by using the power of knowledge and faith, rather than that of authority. Nevertheless, we still consider the cooperation of the authorities both necessary and useful, but inadequate in the face of an unenlightened public whose reactions may therefore be negative.

We believe that the convergence of efforts by science and the State on the one hand, and of the ordinary citizens on the other, is the most effective way to proceed, and invoking the cooperation of all of you, we anticipate that the aims of the symposium currently taking place will be achieved. But at the same time, we are doing everything in our power to raise the consciousness of each and every citizen to the problem of the endangered environment.

We are convinced that you, as preeminent representatives of science, who, in the words of the wise Solomon:

> know the structure of the world and the activity of the elements . . . the alternations of the solstices and the change of the seasons . . . the natures of animals and the temper of the wild beasts, the powers of the spirit and reasoning of human beings, the varieties of plants and the virtues of roots. (Wis. 7.17–20)

will use your knowledge to contribute to our effort. And, for this noble intention on your part, we congratulate you, commend you, and thank you, invoking upon you all the grace and the infinite mercy of God.

Address Before Bulgarian Minister for the Environment, Symposium III, Vidin, Bulgaria, October 23, 1999

THE SUPREME WORD

Welcome to this ship of peace. It is a great pleasure to address to you our own heartfelt greetings as well as those of all our fellow participants, and to welcome you to the third international scientific symposium on the subject Religion, Science, and the Environment. The sub-theme of the symposium this year is "Danube—A River of Life," which is why it is taking place on a ship sailing along the Danube. Our subject is clearly of interest to all countries situated along the Danube, one of which is yours.

Without a doubt, science has important things to say on this subject because it studies and records the actual situation, determines its causes, predicts their effects, and recommends measures to address it. Both the state administration and political leadership have an equally important voice in these matters because it is up to them to decide on the necessary measures and to ensure that they are implemented. Within this whole process of scientific, technical, and administrative activities, it may appear at first glance that the Church and religion have nothing to contribute. In fact, they contribute the most essential element: namely, belief in the moral necessity of these measures, and the moral justification of the entire effort. Moreover, they contribute their blessing and the blessing of God, without which nothing good can be achieved.

We say "the most essential element" because, as we all know, without belief in the effort we are making, without a good "morale" among those who fight, the initial zeal will soon wear off, intensity will slacken, and the aim will not be achieved. And we are all well aware that the goal we are pursuing—namely, to clean up the Danube so that it remains a river of life and does not become a river of death—requires that all people be mobilized. Otherwise, anything that one person tries to build, another tears down. However, mobilization in a struggle whose utility is not obvious to the ordinary citizen may be achieved only if that citizen believes that this is a commandment of God, which it certainly is. And it is the commandment of God because He commanded Adam and Eve—and through them, the whole human race—to work and to protect the Garden of Eden as the earthly environment in which He placed them. So this concept of protection includes safeguarding the environment from all man-made forms of destruction. Unfortunately, we are today witnesses of the daily indifference or even deliberate damage to the environment by our fellow human beings, either individually or collectively.

COMMITMENT AND SOLIDARITY

These are the reasons why we have adopted and launched an effort to sensitize individuals and entire peoples, governments, and nongovernmental organizations to the present cause. We would like to convey to you our commitment to this effort to improve the living standard of your people and of all peoples. We also convey to you the affection and blessing of our Holy Mother, the Great Church of Christ, and mine personally. And we convey to you the message that God is pleased when we, acting out of love, do everything in our power not to burden our neighbor and fellow human beings with our own waste and dirt, which make human life difficult. Such an attitude, if it is generally accepted, implies that we, too, will not be burdened with the wastes of all kinds.

The Danube, like all rivers, is God's gift to humanity. Within His all-wise organization of and dispensation for the world, God offers humankind a wide variety of services. We must not transform this world into a conduit for garbage; and where this has already happened, we must take all the necessary measures to restore the ecological balance.

We should like also to express my satisfaction that this message, a message of genuine solidarity and universal cooperation in the field of peace, has met with good response and acceptance. Rejoicing in this, we shall continue to propagate the same message until it has become generally received and understood, and contributes to improving the entire situation. Once more, please allow me to express to you my pleasure and gratitude for your visit to us on this ship and for your good intentions toward our common effort. We ask that you kindly convey my warmest greetings and our due message to his Beatitude Patriarch Maxim of Sofia and all Bulgaria, as well as to the President and Prime Minister of our dear Republic of Bulgaria as well as to all the Bulgarian people.

Closing Ceremony, Symposium III, Galati, Romania, October, 25, 1999

GRATITUDE FOR EFFORTS IN THE PAST

With the grace of God, we have successfully concluded the third international scientific symposium of the program Religion, Science, and Environment, which focused in detail this year on the specific theme "Danube—A River of Life." In our express desire to sensitize all the peoples living along the banks of the Danube to the urgency of this problem, we have held this symposium on board ship and traveled through all the States along this river. It was a pleasing experience to ascertain that all responsible factors are already aware of the problems and intend to take measures to address them.

We have heard the conclusions and recommendations of the distinguished participants in the symposium as formulated by specialists. We express our profound gratitude to all the eminent delegates, who have shed light on all aspects of the subject in their well-documented papers. We also thank all the authorities, as well as the people of the countries through which we have traveled, for their warm welcome, understanding, and support for our endeavor. In particular, we would like to thank the members of the Religious and Scientific Committee of the symposium, and all those who have contributed to ensuring its excellent proceeding and successful outcome, for their tireless efforts and flawless concern to ensure its smooth flow and harmonious progress, assuming the responsibility of the proceedings on land and on board. We would like as well to congratulate in advance and thank all those who will adopt the conclusions and work toward the implementation of the decisions of this symposium.

We express our joy in the fact that this year's "Halki Ecological Institute" was a product and application of the decisions of our second international symposium, which took place on board a ship two years ago; its particular concern was the Black Sea. We hope that the greater awareness of environmental problems will lead progressively to more and successful measures that aim to preserve ecological equilibrium and natural harmony. The public has almost universally become aware of the fact that a problem exists, and that it is possible to assume measures to deal with the threat and restore the natural balance of our world, which has been so severely disturbed.

DECREES FOR NATURE

Until societies have been sensitized to such a degree that they are themselves able to produce this awareness, thereby keeping alive their sense of responsibility for the protection of the environment and for the assumption and application of the necessary measures, we shall continue to proclaim everywhere the necessity of addressing the environmental problems for the sake of all humanity.

We have all agreed that we are not permitted to hurt our fellow human being. This principle is usually understood and applied with respect to direct harm, for which, in any case, the law provides restitution. Consequently, those members of society who fail sufficiently to respect this principle are obliged by law to comply with any necessary penalties also imposed. *We now need to go one step further and agree even to pass appropriate laws*—at least those who have not already done so—decreeing that indirect damages are also unacceptable. That is to say, those damages are also unacceptable that have been caused by activities far removed from their effect, whose consequences are cumulative, or are manifested in time, so that the cause-effect relation may not be immediately visible. These causes include pollution of all types, which are cumulative and thus frequently cover both long periods and long distances. The more scientific observation consolidates our knowledge about the harmful effects of the responsible, and as a result, the more in number and in efficaciousness will be the measures taken in order to avoid such causes.

Our struggle, then, is not in vain, and our effort is not without benefit. The symposium has offered knowledge and moral sensibility, and therefore has provided the spiritual infrastructure required to assume and apply the necessary measures.

Address at St. Jacob Lutheran Cathedral, Symposium V, Stockholm, June 8, 2003

THE SPIRIT OF RECONCILIATION

We feel great joy in visiting you today, especially on this solemn festival, on which you celebrate the feast of Pentecost, the descent of the Holy Spirit upon the disciples of Christ. May this Holy Spirit—the third person of the Holy Trinity, consubstantial with the other two, who is present everywhere and fills everything, the treasure of blessings and giver of life, who made perfect the Apostles, the guide and inspirer of the saints, who reveals the whole truth to those who accept Him with purity of body and heart—descend into the hearts of all of us, because only if we are guided by Him and we put our trust in Him, can we understand and speak according to God, as the saints of God did in the past.

Historic conditions have separated us from each other and have created the well-known ecclesiastical divisions and the mutually exclusive ecclesial communities, which are not in spiritual communion with each other, in spite of the fact that they believe in the same Lord Jesus Christ and in His saving mission. However, our hearts have not ceased to regard you as brothers in Christ, and to seek the destruction of the walls that separate us, and our rapprochement for which we fervently long.

Moreover, we express our hope that the dialogues presently in progress will draw us closer to each other, and will bring to the surface the real reasons of our separation, reasons that were not always religious or theological, but often the result of a lack of communication for geographical and historical-political reasons. The faith and teaching of the undivided Church during the first centuries always serves as the firm ground for the re-encounter and unity of us all.

LESSONS LEARNED TOGETHER

We have reached the concluding stage and arrived at the final port of our fifth international symposium on Religion, Science, and the

Environment. The participants have together explored the Baltic Sea as a common heritage and a shared responsibility of the surrounding regions. We have been informed by scientists, environmentalists, and politicians about the problems resulting from human abuse and exploitation of the natural environment; and we have heard of ways in which these critical problems are being confronted and resolved.

What we have learned very clearly is that western society has not yet reached the point of sharing the earth's resources or of relating in a responsible and restrained way toward the natural environment. This is painfully evident in the states of the Baltic Sea, where there is such a sharp contrast between the nations in the western and in the eastern parts of the Sea. The discrepancy between the various states in relation to such matters as national income is unjustifiable and unethical. How can our attitudes and actions be reconciled with the spirit of Pentecost?

Nonetheless, it is our fervent prayer and hope that, as we behold before us the water gathered from the Baltic Sea, we may invoke the grace and presence of the Holy Spirit, so that this same Spirit may come upon us, upon the people of this nation, and upon the peoples of this region "as a rush of mighty wind," filling our hearts so that we may hear and speak the tongue of nature, the mother tongue that unites us all.

Honorary Doctorate, University of Amazonas, Symposium VI,
Manaus, Brazil, July 14, 2006

KNOWLEDGE AND LEARNING

It is a wonderful privilege to be presented with a degree *honoris causa* in a part of the world where the sensitive and appropriate use of human knowledge will make such a huge difference to the destiny of mankind. Here in the Amazon region we are continually reminded of the beauty, richness, and complexity of God's creation, and we are conscious that understanding this complexity is an awesome challenge, even for the most brilliant human minds. An even greater challenge lies in deciding how to respond to this complexity, how to harness human knowledge and husband the gifts of God, which are so abundant here in Brazil.

We accept this honor not on our own behalf but on behalf of an ancient institution, the Ecumenical Patriarchate, whose origins go back to the very dawning of the Christian era, a time when people were struggling to hold in balance many different forms of knowledge and understanding.

THE TRADITION OF EDUCATION

Indeed, some of the most heroic moments in that struggle occurred in the fourth century of the Christian era, soon after our Patriarchate was established. An outstanding role was played by three great teachers and hierarchs of the Church whom we regard as the patron saints of literacy and education: St. Basil the Great, St. Gregory the Theologian, and St. John the Golden-Mouthed, or in Greek, *Chrysostomos*.

These three teachers were deeply versed in the formal learning of the classical world, both Greek and Roman. They were trained in the philosophy, science, and literature of the classical age, an age whose achievements are still ranked among the finest accomplishments of the human spirit. Indeed, it is commendable that the languages and culture of ancient Greece and Rome have been studied at a very high level here in Brazil.

But in addition to formal book-learning, the three saintly patrons of Christian education understood something else. They knew that the highest form of knowledge comes through humility, self-discipline, and the contemplation of God. This kind of understanding is entirely accessible, in some ways especially accessible to people who are not educated in the formal sense.

This sort of knowledge—the contemplation of God and His creation—does not require great intellectual or rhetorical feats. If human beings achieve some glimmering of the mystery of God, they do so not by studying books, but by humbling themselves before their Creator and overcoming the vanity and self-love that are a perpetual temptation for all human beings, especially those who are intelligent and well-read. The fathers of the Church respected formal education as an appropriate use of the talents we are given by God, but they warned against the spirit of pride, which can become a trap for those who are endowed with particular intellectual gifts.

In addition to counseling against intellectual arrogance, those early hierarchs were defenders of the poor. They found the courage to criticize those who held power and privilege in the world with no regard for the less fortunate, and they also helped the poor in practical ways by establishing hospitals and distributing food to those who had none. In their writings, the three hierarchs combine intellectual brilliance and zeal for the truth with the compassion and humility which flows from a sense that the greatest mysteries of all, the mysteries of God, can never fully be expressed in human language. This delicate relationship between wisdom of the kind that is pursued in universities, and the divine wisdom, which surpasses all human categories, was also understood by the late Pontiff, His Holiness John Paul II. In November 2004, during the final months of his papacy,[2] he made an important act of reconciliation between Roman Catholic and Orthodox Christian people. Pope John Paul agreed to the transfer, from the Vatican to the Ecumenical Patriarchate, of the earthly remains of two of the three hierarchs, Saints John Chrysostom and Gregory the Theologian. This gesture, made in a spirit of pure goodwill, was profoundly appreciated by Orthodox Christian people, especially those in our home city where both saints had served as Archbishops of Constantinople. In this country, which is home to more Roman Catholics than any other, we salute this hugely important act of reconciliation.

By their teaching and example, the early patrons of Christian education bore witness to an important principle: The pursuit of knowledge is, of course, a noble human enterprise, but we should never forget to ask hard and searching questions about the purpose for which knowledge is being pursued. Are we seeking knowledge for the enduring benefit of mankind and all living things, in all generations? Or is the purpose of pursuing knowledge simply to gain material or perhaps military advantage for a small group of people, even if this advantage is obtained at the expense of the people who are less powerful? In the twentieth century, we saw unprecedented and terrifying evidence of the extremes of good and evil, which the pursuit of knowledge can bring about. On one hand, mankind developed wonderful new ways to relieve suffering by conquering disease and malnutrition; on the other hand, man acquired the ability to destroy every living thing on earth, by the use of devastating weaponry, or simply by the reckless pursuit of economic advantage without regard for the earth's ecological system and its delicate equilibrium.

---

2. Pope John Paul II died on April 2, 2005.

THE LIBRARY OF LIFE

To understand this equilibrium, there is no better place to look than the Brazilian rainforest, which has been called the "library of life" because of its extraordinary variety of animals, trees, and plants, all of them closely and miraculously interconnected. The indigenous peoples of this region are the stewards and guardians not only of the forest itself, but also of a vast store of knowledge about the forest. They know the properties, and potential uses, of every living thing around them. The outside world is jealous of that knowledge, and the peoples of the Amazon are understandably, and justifiably, cautious about sharing it.

It is not the task of the Ecumenical Patriarchate to write laws or contracts, or to pre-judge the outcome of commercial or diplomatic negotiations. But we stand with the people of the Amazon as they consider how, and on what terms, to share their knowledge of the rainforest with the rest of mankind. At the Patriarchate we are perpetually conscious of walking in the footsteps of saintly predecessors who understood, and fought for, the principle that science and learning must be carefully and wisely husbanded, with regard for the all potential beneficiaries, and victims, of that knowledge.

The peoples of the Amazon understand that principle, too; they do not need lectures or sermons from us, but they have our prayers, our understanding, and our unconditional moral support. In the ancient Patriarchates of the Christian world, we have often faced the task of looking after libraries in the usual sense of the word: collections of manuscripts that represent the sum total of human learning and achievement in past eras. We extend our hand in friendship and prayerful support to the peoples of the Amazon, who are worthy guardians and stewards of the library of life.

Homily at the Blessing of the Amazon River, Symposium VI,
July 16, 2006

BLESSING OF THE WATERS

In the Orthodox Church the commemoration of the baptism of our Lord in the waters of the Jordan River constitutes the second most significant feast of the liturgical cycle after the celebration of the Resurrection. The hymns of that day, on January 6, proclaim:

> The nature of waters is sanctified, the earth is blessed, and the heavens are enlightened . . . so that by the elements of creation, and by the angels, and by human beings, by things both visible and invisible, God's most holy name may be glorified.

The implication—at least for Christians—is that Jesus Christ assumed human flesh in order to redeem and sanctify every aspect and detail of this world. This is why, on that day each year, Orthodox Christians will reserve and bottle a portion of the blessed water, with which they subsequently return and bless their homes and families, offices and spaces, gardens and animals.

The breadth and depth, therefore, of the Orthodox cosmic vision implies that humanity is only *one* part of this magnificent epiphany. In this way, the natural environment ceases to be something that we observe objectively and exploit selfishly, instead becoming a celebration of the profound interconnection and essential interdependence of all things, what St. Maximus the Confessor in the seventh century called "a cosmic liturgy." Thus, the future of this planet assumes critical importance for the kingdom of heaven.

THE SPIRITUAL WEB OF LIFE

In blessing the waters of the great Amazon, we proclaim our belief that environmental protection is a profoundly moral and spiritual problem that concerns all of us. The initial and crucial response to the environmental crisis is for each of us to bear personal responsibility for the way that we live and for the values that we treasure and the priorities that we pursue. To persist in the current path of ecological destruction is not only folly. It is a sin against God and creation.

It also constitutes a matter of social and economic justice. As we mentioned in our opening address, there is a close link between the living conditions of the poor or vulnerable and the ecology of the planet. Those of us living in more affluent nations either consume or corrupt far too much of the earth's resources. Conservation and compassion are intimately connected. The web of life is a sacred gift of God—so very precious and so very delicate. We must honor our neighbor and preserve our world with both humility and generosity, in a perspective of frugality and solidarity. The footprint that we leave on our world must become lighter, much lighter.

When we understand the intimate connection and interdependence of all persons and all things in the "cosmic liturgy," then we can begin to resolve issues of ecology and economy. Then our generation would consider the welfare of future generations. There would be a code of ethics to determine behavior and trade, and a clear sense of this world as our common responsibility, with us as its caretakers.

This world was created by a loving God, who is—according to the foremost and traditional symbol of faith in the early Church—"maker of heaven and earth, and of all things visible and invisible." The Judaeo-Christian Scriptures state, in the opening book of Genesis, that "God saw everything that was created good and, indeed, it was very good" (Gen. 1.31). How can one stand before the awesome beauty of the Amazon River without recalling this original plan of God? May God bless this water as He did through His divine presence in the Jordan River. May we all long rejoice in the beauty and sanctification of the waters of this magnificent river.

Closing Ceremony, Symposium VI, Manaus, Brazil, July 20, 2006

"WE ARE ALL IN THE SAME BOAT"

We have, with the grace of God, arrived at the end of our symposium on the Amazon, which has at once been a pilgrimage, a time of learning and reflection, and an opportunity for joyful fellowship with our Brazilian hosts, to whom we express our profound appreciation for their warmth, generosity, and openness of heart.

From the national and state authorities as well as the religious leadership of this country, we have received gracious hospitality and enormous practical support, for which we offer sincere thanks. In the local communities, which we visited, we were warmly welcomed and deeply touched. In our unique encounter with the indigenous peoples of this region, we witnessed and felt their profound sense of the sacredness of creation and of the bonds, which exist between all living things and people. Thanks to them, we understand more deeply that, as creatures of God, we are all in the same boat: *"estamos no mesmo barco!"*

We have listened to some of the most intelligent thinkers in Brazil explaining to us their hopes and concerns for their country and the world. Furthermore, through the wisdom of the indigenous peoples, we have been reminded of our obligation toward future generations. Among the blessings we have enjoyed in Brazil is a new understanding of the ancient images, which our faith uses to describe the delicate relationship between God, man, and creation. Let us reflect briefly on two of these images, namely fire and water. Our faith was born in a region where water is often desperately scarce and where fire spreads easily. In our Mediterranean homeland, it comes naturally to represent the soul's need for communion with God as a longing for water: "My soul thirsts for Thee like a dry and thirsty land" (Ps. 142.6). Elsewhere in Scripture, water is used as a symbol of God's judgment. This is most obvious in the story of the Noah and the Flood (Gen. 6–8). Fire, too, is an element, which in some places describes God's judgment and in others refers to God's greatest gifts, including the gift of the Holy Spirit as it is revealed at Pentecost (Acts 2.1).

These images speak to us powerfully here in the heart of the Amazon, although the natural environment differs greatly from our homeland. We are in a place where water seems abundant and where fire is difficult to spread, at least in places where the forest has remained in its natural state. Nevertheless, we have been informed by Brazilian scientists that this long-established equilibrium may be changing, with fearful consequences. The forest is becoming drier and easier to burn, and the effects of this—so we have been told—may be felt throughout the world, increasing the frequency of extreme weather conditions such as hurricanes and floods. But all is not lost. In visiting local communities nearby, we have observed an environment where water and the natural elements, including fire, can still be experienced as blessings from God.

THE WAY FORWARD

If, by the grace of God, we are able to proceed with our plan for a symposium in the Arctic next year, then we shall see for ourselves how melting snow and rising sea levels could directly threaten the lives and livelihoods of billions of people. Yet the connections we have studied between different parts of the world can also work in a positive way. Continued protection of the river and rainforest, so well preserved in the state of Amazonas, will surely have beneficial effects for the whole earth.

If we are to have any hope of breaking the vicious circle of global environmental destruction, then we shall need an effort, which is ecumenical in the best sense of the word, involving people across religions, races, and continents. In particular, it must embrace leaders and thinkers across disciplines, ranging from theology and religion to biology and economics. We have been blessed during this symposium by a remarkable meeting of minds, through which religion and science were able to share the same language and concerns.

At the mouth of this great Amazon River, there is a city called Belem, a name that ultimately derives from Bethlehem, the birthplace of Jesus Christ. At Christmas time, the Orthodox Church sings of the "Tree of Life," the tree, which stands for perfect communion between God and man, growing anew from the place where Jesus was born. The "Tree of Life" can only grow in conditions where there is neither too much nor too little water. These conditions were granted to us by God but may be so easily destroyed by man. Our fervent prayer is that humanity may walk humbly before our Creator, acknowledging our common origins and destiny with all of created nature. Then, there will be hope for the Tree of Life to blossom for the life of the world and to the glory of God. *Muito obrigado.* God bless you all.

CIVIC AND POLITICAL ASSEMBLIES

Address to the Plenary of the European Parliament, Strasbourg, France, April 19, 1994

UNITY AND COMMONALITY

With much joy and deep satisfaction, we have come among you, on these official premises of the European Parliament, which in people's minds vividly represents the center of renewed historical efforts to achieve European unification.

The unification of Europe, to which you have devoted your efforts as representatives of the will of your peoples, is a familiar task to us. We minister to a tradition of seventeen centuries of caring and struggling for the salvation and unity of European civilization. We, the elder Patriarchate of the Constantinople–New Rome, together with the other European axis, Old Rome, have not been able to make this unity visible. We are most deeply grieved by this fact. We continue, however, to pursue in common our initial witness that a political unity that is separated from civilization, namely without the fundamental meaning of human relationships, cannot lead to the achievement of a united Europe. The unity pursued by the peoples of Europe can only be realized as a unity in the sharing of a common meaning of life, as a common goal of our human relationships.

It is surprising to observe how the realistic and most deeply democratic organization of the Orthodox Christian Church, with its rather large degree of administrative autonomy and the local independence of its episcopates, patriarchates, and autocephalous churches allows them all simultaneously to enjoy a eucharistic unity in faith. Such a model was, in fact, recently legislated by the European Union under the name of this principle of commonality as the most advantageous method for defining its authority.

In spite of the drastic world changes throughout the history of Europe, Old and New Rome continue to remain axes of reference and of unity for the European civilization. We are referring here to the fundamental meaning of unity, not to some ideological alienation of this concept into religious political doctrines, which often leads to the absolutization of nationalistic and racial particularities.

It is our belief that European unification will not be possible if such absolutes dominate. We are aware that, at this very moment, many of you have put before us and before the Orthodox Catholic Church, which we serve in her senior-ranking diocese, the tragic reality of a horrendous war in our times, in which Orthodox populations of Europe have become embroiled, and where there is fighting among neighboring Christians and people of other faiths.

The Ecumenical Patriarchate and the Orthodox Church generally respect the ethnic traditions and sensitivities of all peoples. However, we most categorically condemn every kind of fanaticism, transgression, and use of violence, regardless of where these may originate. Our persistence in the need for free and peaceful communication among peoples, as well as mutual respect and peaceful coexistence among nations remains unmoved, as we also underscored in the "Bosphorus Declaration" during the Conference on Peace and Tolerance recently convened at our initiative.[3]

POLITICS AND POVERTY

You are the prime contributors to European unification. It is your obligation as political leaders, especially given that you are the ones exercising legal authority, to see to the protection of the weak and every kind of minority, the safeguarding of freedom of thought and speech, as well as the movement and residence of persons where their natural, spiritual, and social needs require. In general, it is your obligation to create those kinds of conditions, which would allow for the promotion of cooperation and unity among peoples and nations. In conjunction with this, you have an obligation to reduce, even more, to remove inequality of development, which is evident between the wealthy "developed" world and the "undeveloped." Such inequality endangers the future of humanity and the natural world.

United Europe must not only offer a plan of unified economic development or simply a program of collective defense. With things being as they are, the vision demands a unified social strategy of peaceful and constructive cooperation for all peoples of Europe. It is a question of civilization. It is an issue of understanding interpersonal relationships. It is a matter of acceptance of one another's national traditions.

A SPIRITUALITY OF VULNERABILITY

The Ecumenical Patriarchate of Constantinople–New Rome, which we humbly represent before you today, does not pretend to bring to this center of European unity political strength, economic power, or ideological claims. This is not our mission. Permit us to note, however, that our

---

3. See the first volume in this series, entitled *In the World, Yet Not of the World* (New York: Fordham University Press, 2010).

experience through the centuries paradoxically confirms that all power, which continues to stimulate history, is "made perfect in weakness" (Cor. 12.9). We submit to you the experience of our recent efforts. Whenever we have made any efforts, based on our power or, more properly, beyond our power, out of our concern for ecumenical unity of the Christian churches, the fruits that we were deemed worthy of receiving were in fact the product of our frailty and not of our force. In 1920, the Ecumenical Patriarchate, on its own initiative, addressed a formal encyclical to the entire world in order to convene a kind of league of churches modeled after the "League of Nations," the forerunner of today's United Nations. Through this initiative, and with the cooperation of the Protestant confessions, the World Council of Churches was eventually founded.

A similar experience is derived from an equally significant initiative of the Ecumenical Patriarchate to inaugurate, along with its other sister Orthodox Churches, bilateral theological dialogues with the ancient Oriental churches, as well as the Roman Catholic, the Old Catholic, the Anglican, the Lutheran, and the Reformed churches. Perhaps, the more senior members among you will remember the historical meeting of our predecessor, Patriarch Athenagoras, with Pope Paul VI, both now of blessed memory, in Jerusalem in 1964. This was the first such meeting of the primates of Old and New Rome since the Great Schism of 1054. You may further recall the historical mutual lifting of the anathemas between these two churches in 1965 and the exchange of visits between Pope John Paul II and our immediate predecessor, the late Ecumenical Patriarch Dimitrios.

We ourselves continue these efforts. Recently, we extended them by attempting an interreligious rapprochement. We therefore convened an international, inter-faith conference, under the aegis of the Ecumenical Patriarchate, on the theme "Peace and Tolerance." We are fully aware that the cultivation of a peaceful climate for coexistence and creative co-operation—both among religions and churches as well as among national states, races and traditions—demands radical change. Dialogue, international conferences, communication among leaders responsible for drafting legislation, growing closer through goodwill, and abandoning the notion of irreconcilable differences: All these are positive and useful steps, yet they are not enough. The problems of the contemporary world in general, and the problems confronting Europe in particular, demand fundamental reevaluations of our cultural choices. In other words, they presuppose a new cultural model.

THE CRITICAL ISSUE OF UNEMPLOYMENT

Two emphatic paradigms bear special witness to this need. The first is the tragedy of unemployment, which plagues Europe today. It is obvious that neither moral counsel nor fragmented measures of socio-economic policies would suffice to confront the rising unemployment. Indeed, the problem of unemployment compels us to reexamine the self-evident priorities of our society, such as the absolute priority of so-called "development," which is measured solely in economic terms. We ate inexorably trapped within a tyrannical need continually to increase productivity and, as a consequence, continually to create newer and larger quantities of consumer goods. Placing these two necessities—of development and of economy—on an equal footing imposes the constant need for greater perfection of the means of productivity, even while continuously restricting the power itself of production, namely the human potential. Concurrently, consumer needs of this very same human potential must constantly increase and expand. Thus, the economy becomes independent of the needs of society, functioning without human intervention and developing into an industrial method, which tries to equate abstract proportions.

Perhaps, as a result of the acute problem of unemployment, it is now time, instead of concerning ourselves with the self-centered demands for our individual rights, to prioritize personal productivity within the context of human relationships. Civil management of public affairs must answer the following questions: Who or what will inspire the European of today to give priority to inter-personal relationships? What will be the political mandate that will convince humankind willingly and joyfully to sacrifice its impetuous thirst for consumption and its limitless demand for unquenchable productivity in order to rediscover the communion of life within the community of persons?

Politics play a critical role in the radical changes in our understanding of human life. However, in people's conscience, such changes are consolidated only by the persuasion of experiences conveyed through religious traditions. If the classic and renowned studies by Max Weber, Werner Sombart, and R. H. Tawney hold true in their findings, then, at the base of contemporary European understanding of work and economics, one may find a concrete receptivity to Christian theology. If this finding holds true, then a new concept of understanding work and economy will inevitably pass through some theological revision. The circumvention of theology by various ideologies has not proved convincing that the latter might bring about any realistic solutions. Behind the modern impasse of European life hides a theological position.

THE RADICAL ISSUE OF ECOLOGY

We believe that similar conclusions may be derived in the second issue, which is equally critical and distressing in our times, namely the problem of ecology. All of us are aware of the nightmarish proportions of this problem as it increases day by day. Permit us to hold to our conviction that the ecological problem of our times demands a radical re-evaluation of our understanding of how we see the entire world; it demands a different interpretation of matter and the world, another perception of the attitude of humankind toward nature, and a new understanding of how we acquire and make use of our material goods.

Within the measure of our spiritual capacity, the Orthodox Church and Orthodox theology endeavor to contribute to the necessary dialogue concerning this problem. Thus, at the initiative of the Ecumenical Patriarchate, the Orthodox have established September 1st of each year as a day of meditation and prayer in order to confront the continuing ecological destruction of our planet. Having convened an international conference in Crete, we have further inaugurated a systematic theological study of this problem.

BEING AND ACTING IN SOLIDARITY

However, our efforts will be meaningless if they remain fragmented. Therefore, taking advantage of the fact that we are standing here before you, we hasten also to declare that we are prepared to place our modest efforts at the disposal of the European Parliament for any future study and concern of a pan-European response to the ecological problem. Permit us to declare the same readiness in reference to the aforementioned problem of unemployment plaguing Europe.

Mr. President, Ladies and Gentlemen, Members of the European Parliament, your gracious invitation has permitted us to share this limited but valuable time in order to communicate personally with you. We feel the overwhelming burden of our responsibility. With our humble words, we have endeavored to review the history and the experience of an institution, which for seventeen centuries continues to function as an axis for the unity of European civilization. We aspire to continue the tradition of the Ecumenical Patriarchate of Constantinople–New Rome and to continue to preach the Word of God as did the Patriarchs of Constantinople John Chrysostom,[4] Gregory the Theologian,[5] Photius the Great,[6] and a myriad of others, giants not only in the realm of ecclesiastical history, but also of recognized stature in European history as well.

Historical conditions have drastically changed the contemporary world scene. Please accept our presence here as a simple reminder. We remind you that we exist. And we continue to minister and to bear witness to the common struggle in our care for understanding and hope worldwide. The metropolitan sees of the Ecumenical Patriarchate throughout all of the European countries, the hundreds of parishes of Orthodox faithful in Central and Western Europe, immigrant and local populations, constitute our flock and are also people of your political constituencies. From outside of the boundaries of the twelve-nation European Union, a great number of other populated nations belonging to the Orthodox ecclesiastical tradition are also following the same European journey. Permit us to express the hope that these peoples will be called soon to participate in the life and institutions of a united Europe.

Through its faithful adherents and under the developing circumstances, the Ecumenical Patriarchate continues, in its ecumenical mission, to serve as an essential part of this European dimension. Beyond the ideological orientation of each of you, beyond each individual's personal or metaphysical conviction, we kindly ask that you accept the readiness of the Ecumenical Patriarchate to support you in your efforts for European unification, for a Europe, which will not exist for itself alone, but for the benefit of all the world.

---

4. St. John (349–407) was Archbishop of Constantinople from 398 to 407.

5. St. Gregory (328–389) was Archbishop of Constantinople from 380 to 383.

6. St. Photius (810–893) was Archbishop of Constantinople from 858 to 867 and from 878 to 886.

Address at United Nations Luncheon, New York, October 27, 1997

UNITED AS NATIONS

As leaders of the international community, you have an awesome responsibility to ensure peace throughout the world.[7] As the successor to the Apostle Andrew, the First-Called disciple of Jesus Christ, we too have a similar responsibility. Our Modesty is charged by Christ to preach the message of peace, hope, and love. We do this, recognizing that our message must be set within the realistic setting of people's lives. We are committed to the universal cause of freedom, religious and political self-determination, and justice.

We have spoken today with the Secretary General of the United Nations, the Hon. Kofi Annan. We thank him for his warm welcome. We look forward to working with him on all issues that touch upon the fundamental message of our Lord's commandment to love one another. We represent the 300 million communicants of the world's Orthodox Christians who span the planet's many continents.

Certainly, the United Nations has recognized the increasing importance of religious communities, as partners in the conflicts that have marked the post-Cold War world. It is a tragic fact that religion has contributed to cycles of violence and fragmentation between and within nation-states, before and since the end of the bipolar international order. However, it is also true that the end of bipolarity has produced new social and political circumstances. Recently, religious entities have directly cooperated with temporal powers in developing principles, language, and concrete policies premised on the rejection of categories of power, which lead to separation, exclusion, disintegration, and conflict.

During the mounting chaos that was part of the implosion of Albania during much of 1997, the Orthodox Church of Albania was the first voice in civil society to call upon all citizens to refrain from acts of violence. Since 1991, the country's Orthodox leadership has worked steadily and closely with its Muslim counterparts.

In Turkey, the Ecumenical Patriarchate has led the way in its efforts to promote interfaith tolerance, by convening the conference on Peace and Tolerance amongst Christians, Jews, and Muslims. This forum produced

---

7. Le Cirque Restaurant.

the "Bosphorus Declaration," whose signatories condemned all religious violence and proclaimed that crimes in the name of religion are crimes against all religion.[8] Similarly, only recently, the Greek Orthodox Church in America inaugurated an official dialogue with Muslim leaders in America. On a broader scale, the International Orthodox Christian Charities (IOCC) is a transnational NGO.[9] This organization, very dear to our hearts, is vigorously opposed to proselytism. It is committed to interfaith cooperation, via philanthropic and welfare projects, for improving literacy and life expectancy, as well as for combating infant mortality and disease.

UNITED AS PEOPLES

The Ecumenical Patriarchate has worked to promote an awareness of the global ramifications of environmental issues. We see this divinely inspired work as integral to the goal of world peace. We shall continue our yearly seminars on environmental issue on Halki and our initiatives elsewhere.

We mention these positive examples of the Orthodox Church's involvement in worldly affairs in order to underscore the fact that our faith offers a rich set of resources useful in promoting the United Nation's noble agenda of conflict transformation and peace building. Unfortunately, the unique resources of the Orthodox Church remain relatively unknown to, and therefore, underutilized by the United Nations. This situation has led to certain situations where Orthodoxy has not had the support it might have had, support that would have assisted the Church's message of love to prevail in the face of historical animosities.

As the Mother Church of Orthodoxy, the Ecumenical Patriarchate of Constantinople is ready to expand all its efforts to rebuild the moral and ethical well being of long-suffering peoples everywhere. We say this without any intention of transgressing the fundamental rights of spiritual self-determination. However, we would hope that the West might also make an effort to understand the unique hardships that the Orthodox Christian East has suffered during the long decades of persecution. We ask that non-coercive aid programs be extended to Orthodox Christians in areas where they are recovering from decades of repression and persecution.

---

8. See the first volume in this series, entitled *In the World, Yet Not of the World* (New York: Fordham University Press, 2010).

9. See www.iocc.org.

We believe that Orthodoxy's rich creation theology rests on the assumption that the entire cosmos is an integrated whole. The Orthodox Church's theological and existential goal of the integrity of all creation is consonant with the United Nation's goal of peace building. Orthodoxy's understanding of the human being as person, as a microcosm of the cosmos, assumes that our humanity is existentially meaningful only through the free and conscious engagement in relation with others. The Ecumenical Patriarchate is committed to transforming the human condition. Our vision of freedom is relational, and it is consistent with UN efforts at transforming post-conflict situations, by restoring the torn fabric of individual and community life.

The Orthodox Church transcends linguistic, ethnic, and national divisions. Our holy Orthodox Church is modeled on the Trinitarian principle of unity-in-diversity, whereby heterogeneity and uniqueness are fundamental aspects of our humanity. The Church's experience is comparable to the UN itself, an entity whose unity is the result of the diversity of its membership.

Dear coworkers in the vineyard of peace, our presence here today is meant to reaffirm the commitment of the Holy Orthodox Church to the UN agenda. We humbly serve notice that our Church is capable of offering much to better the health of the UN family.

We exhort you to assume the responsibility, which has been given to all of us by God, our Creator, to renew collectively our commitment to restoring peace, justice, and the integrity of all creation. We ask you to consider the creative gifts of the Orthodox Christian community as a resource for change. In this respect, we respectfully offer you the spiritual resources of the Ecumenical Patriarchate.

Remarks During Luncheon by the Minister of Environment, Oslo,
June 12, 2002

SCIENCE AND THEOLOGY

The primary purpose of our interest for the protection of the environment is our concern for humanity in our own time and in future generations. Of course, we are not indifferent toward the preservation of natural elements that are endangered. Indeed, we see in them God's love and wisdom. Therefore, out of respect for God, we consider it a duty of our love toward Him to preserve His creation, which bears witness to His goodness.

Our attitude toward the whole of creation is influenced by our faith in God and our love toward Him and His works, and especially toward our fellow human beings. We see the entire world as an expression of the goodness that characterizes the Supreme Being. We know that everything that exists has a reason for its existence. Nevertheless, we also believe that the original harmony of every being in the universe has been disrupted through the intervention of the human will, which has rebelled against it. The only way in which a complete harmony can exist in accordance with the original divine plan is if the human will embraces and voluntarily submits to this plan.

Science concentrates all of its attention on one object. It is of necessity, analytical. Therefore, the gathering of general principles that regulate the universe—which many of the Church Fathers, and especially St. Maximus the Confessor, describe as the "inner principles"—constitutes a philosophical task. In the theology of the Church, which delves into the revealed truth and appreciates human knowledge, this is described as "the heavens declare the glory of God." Consequently, any effective resolution to the contemporary environmental crisis requires a theological basis, whose natural result is an appropriate environmental ethos.

8

Interviews and Comments

A Selection

CHARLIE ROSE

CHARLIE ROSE: Welcome to the broadcast.[1] Tonight, a conversation with one of the most important religious figures in the world. He is His All Holiness, the Ecumenical Patriarch Bartholomew of Constantinople. He's the leader of 300 million Orthodox Christians. We spoke in Atlanta during his current visit to the United States. Tomorrow in Washington he meets with President Obama. From the beginning he has been on a mission to modernize the church and make it more relevant. Early on he became identified with environmentalism by incorporating it into his spiritual message. He has preached in the spirit of dialogue and understanding among all religions.

HIS ALL HOLINESS ECUMENICAL PATRIARCH BARTHOLOMEW: I came to the United States for our eighth international interfaith and interdisciplinary environmental symposium, which took place last week in New Orleans. We have this series of ecological symposia in order to create sensitivity and awareness among our own faithful. But there is more to it than that. This is a real concern of the church as it is and must be a concern of all of the humankind, human beings, because we see everyday more and more the dangers and the threats of climate change, of pollution, and so on. And we have to create this awareness and sensitivity so that at least our children and grandchildren may live in a more human, more beautiful, safer, and cleaner world. Usually we speak about the education of our children and the good food of

---

1. This interview with Charlie Rose was part of a one-hour program, which aired on November 2, 2009.

our children, but what about the air that they breathe and the water they drink? Now, and tomorrow, and after tomorrow, we have to think of the coming generations, our posterity. That is a duty of the church and that is why the Ecumenical Patriarchate initiated this symposium and environmental activity.

C.R.: Throughout the world. I think it was Al Gore but it may have been someone else who first called you the "Green Patriarch." Do you accept that?

H.A.H.: Thank you. This is an honor for me. It is recognition, if you like, of our humble environmental activities. It is good simply for a broader dissemination of this environmental message.

C.R.: Your message is spiritual, that it is a spiritual responsibility.

H.A.H.: Precisely. It is important for the church and its leaders to connect the care for our planet to a spiritual vision and life. I used to say that the ecological problem is not primarily an economic or a political problem. It is predominantly a spiritual and ethical question, because it shows our relation not only to God or with God, but also our relations with the Creator of everything according to our faith, but also our relation to his creation, because the creation is creation of God is also sacred, and we have to respect it and to protect it according to the order of the first book of the Old Testament, and simply to use it and not to abuse it. We must think of the coming generations who must share the same treasures, the same resources of the planet, as we do today.

C.R.: So how do you plan to use your influence to make a difference?

H.A.H.: Principally by creating a healthy consciousness of our responsibility, for respect and protection of the natural environment of our planet, of which we are simply stewards and not proprietors.

A Common Vision

We are pleased to welcome you here, at the very heart of Orthodox Christianity.[2] You have taken time out of your schedule to visit us and this

---

2. The material in this section was taken from a seminar of Greek, Turkish, and German journalists sponsored by the Konrad Adenauer Foundation (October 25, 1996). This address to the media is included here as an introduction to this section, which contains excerpts from interviews and brief remarks by the Ecumenical Patriarch. Although there is little direct reference to the environment, it nevertheless reveals the openness and readiness of the Ecumenical Patriarch to work with people of all disciplines and experience in confronting ecological issues.

venerable and historic Ecumenical Patriarchate and we are honored by your presence. Here, in the Great city of Istanbul, the meeting place of Europe and Asia, people from all walks of life and from every corner of the world gather to enjoy the mystery, the history, and the majesty of this city. Although for some of you this may be your first time in Turkey, you surely are not strangers to the legacy given the world by the diverse civilizations and cultures that have blessed these sacred lands.

Here, you are standing at the crossroads of civilizations. Look in any direction and you will see the descendants of some of the greatest: the Greek, the Ottoman, the Armenian, the Jewish, the Leventine. Open your minds and hearts and you will be inspired by some of the greatest world faiths: Christianity, Islam, and Judaism, which for centuries have coexisted, and which, at the threshold of the third millennium, are determined to coexist here for many more.

Among these great world religions is our own Orthodox faith, which has been rooted here since apostolic times when St. Andrew, the First Called by our Lord, brought the Gospel of Jesus Christ to these seven hills and founded the Church on this hallowed soil. The Holy Great Church of Christ, now housed in these humble but friendly surroundings, remains the beacon of Orthodoxy and the Mother Church to Orthodox Christians throughout the world. The Ecumenical Patriarchate continues to be held in special honor and today opens her loving arms to welcome you to her bosom.

Gathered here today are four—you might say—"unlikely" groups: Turkish journalists, Greek journalists, German journalists, and the Ecumenical Patriarchate. Of course, as journalists you all have your industry in common, but your viewpoints often vary. Sometimes they agree; sometimes they cross each other at points; sometimes they are maverick and shoot off into different directions. For serious journalists, such as those of this distinguished body, their main objective each time, is *not* sensationalism, *not* to flirt with politics, *not* to undermine peace, but to print the truth.

And just how do we fit in? We are a religious institution, a Church. We are in the world but not of the world; and, as such, we too are bearers of truth and are obliged to communicate it to all humankind. So, there are common threads uniting us, but are there enough to support the fabric? Obviously, we all believe there are, because we have all gathered here today under this common roof, a place symbolizing truth, love, dialogue, and understanding: You, as professional journalists with a mandate to report life as it develops from day to day in the world around you; we, as men of God also with a mandate bearing a message of salvation and eternal life intended for the world around us. Each from their own vantage point agrees that their message of truth must not be captive, but must reach our audience—for you, your readership, for us, our faithful believers.

THE HOUR OF ORTHODOXY

Our age is the age of Orthodox witness.[3] The hour of Orthodoxy has come. Today, orthodoxy is respected and welcomed more so than at any other time in the past. The responsibility, then, of its Church leaders is greater than ever before. Wherever western Christianity and western civilization have left a vacuum, Orthodoxy is called to fulfill and to transfigure the face of the earth with its renewed freshness, like the youthfulness of an eagle (Ps. 103.5), with its spirituality and apophaticism,[4] and in general with its inexhaustible spiritual treasures, which belong to the whole world and which for this reason must be spread as widely as possible.

The contemporary and critical issue of the environment justifiably concerned our predecessor of blessed memory, Patriarch Dimitrios. We, too, cannot but continue our interest in this matter. The Ecumenical Patriarchate always endeavors to show its presence in people's lives of every period and to express an interest in human problems and needs in order to assist and support them as it should. Only a few days ago, the Ecumenical Patriarchate convened a special Inter-Orthodox Conference on environmental studies at the Orthodox Academy of Crete, chaired by the distinguished Hierarch and University Professor, Metropolitan John of Pergamon. Furthermore, next year, the Duke of Edinburgh will visit the Phanar in his capacity as Chairman of the Worldwide Wildlife Fund, and

---

3. The text in "A Common Vision" is from an interview with Nikolaos Manginas for the journal *Eikones* (January 24, 1992).

4. The emphasis on knowing and describing God through negation, based on the assumption that God's essence always remains incomprehensible and ineffable.

we shall then surely have occasion to exchange views and initiate plans for the protection of the natural environment from human exploitation, which also implies the protection of human life itself.

ENVIRONMENTAL PROTECTION IS CRITICAL

The Ecumenical Patriarchate is indeed concerned with the matter of the environment.[5] Our predecessor, Patriarch Dimitrios, began by establishing September 1st of each year as a day for intercessions and prayers for the environment. The Ecumenical Patriarchate organizes various ecological conferences, while on a personal level I have begun working with Prince Philip (Duke of Edinburgh) in his capacity as Chairman of the World-wide Wildlife Fund. Next September, on the occasion of the celebrations for the 1900th anniversary of the Revelation of St. John, we shall again organize an International Ecological Symposium. Furthermore, I am to travel to Japan and England next April in order to speak on behalf of all Christians at an international environmental congress. It is our hope that these endeavors by the Mother Church, which have also been adopted by other sister Orthodox Churches during the meeting of the Primates here (in Constantinople, 1992), will bear fruit and mobilize other Christian Churches as well, in addition to the faithful of the Orthodox Church, so that each person—in their own place and their own domain—may contribute in this direction of the preservation of God's creation, the natural environment.

A RELIGIOUS OBLIGATION FOR ALL

As a result of the first five environmental Summer Seminars on Halki (1994–1999) we have established that the protection of the environment in which humanity lives is a divine commandment.[6] Our position is founded upon God's commandment to those whom he first created, that

---

5. The thoughts offered in this section are from an interview with the managing editor of *Oikonomikos Tachydromos*, Yiannis Marinos (January 31, 1995). Also appeared in a book entitled *Orthodoxy and Contemporary Issues* published by Eptalofos, Athens, 1995.

6. This section, "A Religious Obligation for All," is from the foreword for the publication of the Halki Seminar Proceedings (March 15, 2002).

they, according to the teaching of our faith, "labor and tend the garden" in which they were first placed by Him. This is the theological basis for the humankind's role in the protection of the environment. Thus, at the beginning of this new millennium, it is not only our divine obligation to labor and utilize the fruits of His gifts on this terrestrial globe, but also, as His most humble servants, to assume responsibility for tending the Garden of Eden.

Keeping this garden in its proper and pristine condition, and then preserving and protecting it after the dispersions of peoples all over the earth, is also our respectful duty to our fellow human beings, as well as to future generations. For, the pollution or destruction of even a single element of the environment brings hardship on the life of another person. Consequently, there is an ethical responsibility on our part that we not make life difficult for our fellow human beings.

Herodotus tells us that among certain ancient peoples there was a well-known custom that they not pollute the rivers, which they believed to be in the divine order of things. Certainly, conveyed here, there is an ethical condemnation of the hardship brought upon the lives of one's fellow human beings using the flowing waters at some point further downstream. Also well known is the commandment of Moses to the Israelites dwelling in the desert that they follow the Scriptural prescription by taking certain measures to conceal and bury the polluted wastes and keep their environment clean and healthy.

Today, then, we are able to say that the Christian, Jewish, and Muslim religions (the latter of which accepts in part of the Old Testament as encompassing the declaration of God's will to humankind) oblige us to emphasize to our faithful that tending and protecting the earthly environment that we inhabit is a commandment of God; as such, environmental awareness and action are a religious obligation.

An Initiative, Not a Novelty

The Orthodox Church has always embraced the entire creation and indeed prays daily "for favorable winds, the abundance of the fruits of the earth, for those who labor and those who are sick, for those who are traveling by sea or road or air" and, in general, "for the unity of all in Christ."[7] This is a message of reconciliation with God not only of the estranged Man but also of the rebellious nature, which "sighs and suffers with us," desiring to be liberated from subjection to decay (see Rom. 8.22). Therefore, it should not be considered a novelty that we are asking our fellow human beings not to despoil nature, on behalf of which we are praying daily to God. However, it is not true that we assign the highest priority to ecological questions alone. We are simply giving these questions their proper place among many other, equally or even more serious, problems to which we call the attention of our fellow men. Such problems deserving our close attention are peaceful coexistence and collaboration, reconciliation, support for the weak and the needy, respect for the human person, the correct practice of our faith, the return of all things to God, and so on.

ORTHODOX SPIRITUALITY AS MAXIMALIST

The teaching about the creation of the world and the doctrine in general about the natural environment provide the overall parameters within which anyone interested is called to study all the contemporary "ecological" or "environmental" problems.[8] This is why it would constitute an unbearable and unjust "minimalism" if one were to judge the relative stance of the Orthodox Church solely from the perspective of an occasional document at a particular event or from the establishment of September 1st of each year as the day of prayer for the whole creation and environment. Therefore, the entire lifestyle of the Orthodox faithful at every moment of every day is indicative of the sacredness that is reserved for everything created. By way of reminder, it is characteristic that the monks of Mt. Athos will make the sign of the cross in a gesture of gratitude before even drinking a mere glass of water.

---

7. The material in this section is from an interview for the student magazine of Columbia University (March 3, 1998).

8. The interview/message provided in this section was broadcast by the British independent company Granada Television, in its series *This Sunday* on the occasion of Orthodox Easter (January 18, 1993).

A New Worldview

We are obliged to repeat the fact that we cannot save the natural environment with the same methodology and "philosophy" concerning nature
with which we have destroyed it.[9] We require another, different worldview, a new perception of matter and the world. And in this discernment
of a new perception and meaning, it is our conviction that our religious
tradition has an active role to play.

The Heresy of Anthropocentrism

Orthodox theology and spirituality have never deviated toward anthropomonism, which would risk leading to Monophysitism.[10] On the contrary,
it always regards the human person as the center of the whole creation,
the sacredness of which is manifested and cannot be undermined by the
central position and privilege, which humanity has as the image of God
in nature. The ecological problem should not simply be considered as an
"environmental" crisis, because the source of the crisis and the primary
cause of the problem is the one who "embraces the environment," namely
humanity and its darkened conscience.

Healthy Environment—Healthy People

The Orthodox Church and in particular the Ecumenical Patriarchate are
profoundly concerned about environmental issues, inasmuch as they concern the health of human beings and the future of our planet.[11] In any
case, it is our sacred obligation, commanded by God, as faithful stewards
and managers of creation, to protect and not to destroy nature by exploiting it selfishly. In order to mobilize our young people especially, we are
organizing ecological conferences such as the one to be held in September

---

9. The material excerpted in this section is from an interview for Greek television,
TV100 (July 5, 1994).

10. Ecumenical Patriarch Bartholomew offered the thoughts in this section during an
interview with the journal *Helsinki Orthodox Parish* (1994).

11. The comments in this section are taken from an interview with Heikki Tervonen
for the Finnish newspaper *Kotimaa* (May 1995).

of 1995 in conjunction with the festivities for the Revelation of St. John, the futuristic biblical book *par excellence.*

THE HIERARCHY OF VALUES

Our environmental activities touch in a very fundamental way on the sacredness of creation, as well as on the significance and preservation not only of the environment as a context but, by extension, of life and the perfection of all creatures.[12] It is, therefore, a matter of a mission that in and of itself is necessary and which must accordingly be directed to contemporary and critical concerns.

Nevertheless, what we consider to be supremely important in this regard is the transcendence of the arrogant and at the same time self-destructive concept of an economic development without boundaries, which transforms the means into an end in itself. We also regard it as people's obligation to adopt a hierarchy of values, which would serve the meaning of creation and preserve the necessary and sufficient conditions for such a way of life.

ARROGANCE TOWARD NATURE

What prompted the Ecumenical Patriarchate to become deeply interested and involved in working for the preservation of the environment was that we recognized the great dangers resulting from the incalculable and arrogant abuse of nature by humanity.[13] Human beings must remember that they are protectors, managers, and stewards of material creation, and not possessors or dominators over it. By means of various actions, the Ecumenical Patriarchate wishes to convey this message widely and thereby anticipate greater problems awaiting all of humanity and all of God's creation. And we shall continue our efforts out of respect for creation as the work of a divine Creator. We are called together to protect the sacredness and beauty of God's creation.

---

12. The material in this section is from an interview with the journalist Sophia Tsiligianni (May 28, 1995).

13. The material in "Arrogance Toward Nature" comes from an interview with Edmund Doogue, coeditor of *Ecumenical News International* (October 13, 1995).

Sounding the Alarm of Danger

The interest of the Ecumenical Patriarchate for matters concerning the protection of the environment, as for all contemporary issues, is taken for granted.[14] All of us here at the Phanar believe that the burning issue of the environment must be addressed at its root. And the root of this problem, just as the root of so many other problems, is humanity. Human beings exploit their identity as the only rational beings and externalize their selfish attitudes, inflicting significant and incorrigible damage on nature. You see, we are given the opportunity to use creation, but instead we have preferred to abuse it.

As the Orthodox Christian Church, we must sound the alarm of danger. We must work and walk with all those persons who see the great risk and contribute to the restraint of this evil. We, too, must contribute as a Church by raising the awareness and awakening the conscience of all those who remain indifferent. I am certain that, when humanity in its entirety becomes truly conscious of the fact that its existence depends on the environment, the ecological problem will disappear. However, the world must be mobilized now. Appropriate measures must be taken in timely fashion because we have already delayed somewhat. Should we delay still further, then the dangers for humanity will become greater and we shall no longer be able to turn around the current of things.

An Ecumenical Vision

From the moment of its foundation to the end of the ages, the purpose and perspective of the Ecumenical Patriarchate's contribution and service to humanity remain constant: the proclamation of truth, the teaching of the Gospel, the coordination of and ministry to the Orthodox Churches, the sanctification of the faithful, and the transfiguration of the whole world.[15] Within these general aims and fundamental perspectives are included the more particular goals, such as the insistence on tradition, the

---

14. The thoughts presented here are from an interview with Alexi Tsolakis and Panagiotis Christofilopoulos in the educational journal *Athinaios* (January 1996).

15. The material in this section was taken from an interview for the journal *Aerodromics* of Air Greece Airlines (October 9, 1999).

evasion of secularism, and the mobilization of all people in regard to the environment.

The Ecumenical Patriarchate has taken the initiative to remind all Christians that the protection of the environment is a divine command and a human obligation. For the destruction of the natural environment bears dangerous consequences for all of humanity. So the Ecumenical Patriarchate seeks to sensitize everyone because even the ecological destruction is often the result of the small contribution of each person.

A COLLECTIVE CONCERN

The preservation of the natural environment is not an isolated matter of interest to many or few individuals, but rather of collective concern for all persons.[16] This is why we are trying to make as many people as possible aware of the dangers and conscious of the necessity to protect our environment.

THE OBLIGATION TO PRESERVE

The message conveyed by the Ecumenical Patriarchate through its various environmental initiatives is one: that nature was created for the service of humanity as well as for the glory of God.[17] It must not be either used or abused with greed or indifference. We are obliged to preserve this gift of God also for the generations to come.

AN INITIATIVE TO EMULATE

People outside the Orthodox Church regard the contribution of the Ecumenical Patriarchate in environmental matters as very positive.[18] When

---

16. The brief comments here are from an interview on Greek television *Antenna* (November 18, 1999).

17. The brief comments here are from an interview for the Greek journal *Gaiorama* (March 26, 1999).

18. This material from an interview with Grigorios Troufakos was published in the Greek daily newspaper *Eleftherotypia* (Easter, 1995).

Prince Philip visited the Phanar in 1992, he underlined and commended the initiatives of the Patriarchate in contemporary ecological problems, observing that it is the first Christian Church to do so.

SPIRITUAL ROOTS OF A MATERIAL PROBLEM

What people refer to as "genuine human problems" have a spiritual root and cause, stemming as they do from the human heart and soul.[19] Consequently, in order to solve these problems satisfactorily, one requires a change of lifestyle, a conversion of mindset and attitudes of heart—what in ecclesiastical terminology is called *metanoia* (or repentance). So the Orthodox Church attributes great value to spirituality, acting precisely as a good physician who does not seek to heal an illness by eliminating the particular symptoms but by addressing the causes of the disease.

All contemporary "real" problems have a spiritual cause, and the Orthodox Church is right in focusing its attention on the spiritual dimension. At any rate, it is not the first to do this because even the historian Thucydides claimed—many centuries before Christ—that it is impossible for any change to occur in society and in social matters without a change first of all in individuals. And, of course, a change in the character of individuals comes about only through a change in their spirituality. This change is wrought by the wealth of Orthodox worship. For without the grace of God, which is invoked through the liturgy, no human change for the better is possible.

HUNGER AND POVERTY

The problems of hunger and poverty have as their root the human lifestyle and not the insufficiency of material goods to support human needs.[20] It is well known that vast amounts are expended for the production of destructive military weapons and materials. It is also well known that a vast

---

19. Ecumenical Patriarch Bartholomew provided the material in this section in an interview with Prof. Bruno Forto for Italian Television (January 6, 1999).

20. This section provides the text of an interview on Greek radio *Sky 100.4* (April 9, 1999).

majority of wealth and resources are concentrated in the hands of few, as we recently stated in Davos. If these amounts were used for peaceful and resourceful purposes, they would be able to transform our planet into an earthly paradise.

Of course, we are not political leaders in order to propose or impose solutions. Yet we are obliged in the name of our faith and of truth to proclaim the need to change people's lifestyles and attitudes, to preach that which in ecclesiastical terms is called *metanoia* (or repentance), in order for human conditions to improve. The word "repentance" is misunderstood today, calling to mind a sense of guilt for sins that some people consider unessential. By "repentance," however, we imply those things that are more important than the transgression of law: namely, discernment and mercy, justice and compassion.

The lack of a sense of justice leads to greed, domination, the exploitation of the weaker by the more powerful, the abundance of wealth for the strong, and the extreme poverty of the weak. The lack of a spirit of compassion renders the soul indifferent to another person's pain and prevents the development of those things that kindle a sense of justice. Therefore, in proclaiming a change of attitude, we are offering a kindly service to humanity and indicating a way of solving problems of poverty and hunger. Nevertheless, we are not naively optimistic.

At the same time, there are numerous signs that a significant—and, it is our hope, a growing—portion of human societies is conscious of this direction, although we are not ignorant of the fact that the abundantly wealthy minority will continue to increase in wealth. However, as a religious leader, and especially as a leader of the Orthodox Church—the Church of love, justice, compassion, and service—we have no other way but that of proclamation and persuasion.

Our efforts in particular for the protection of the natural environment must also be intensified. Yet, we must broaden the notion of the natural environment to include the human and cultural environment. For it would be a paradox to be concerned solely for the natural environment, and yet be lacking in interest for humanity and our cultural heritage. Our human environment also deserves our love, just as our natural environment deserves our respect. Even more so, our cultural heritage is a monument of the human journey that is deserving of our respect and protection.

A VISION OF RECONCILIATION

The vision of the Ecumenical Patriarchate for the realization of peace on earth and good will among people is a vision of reconciliation, of peaceful cooperation, of respect for the human person and for the natural environment.[21]

ECOLOGY AND MONASTICISM

Inspired as we are by the will to love one another, we wholeheartedly embrace the concern for the natural environment as a way of overcoming or at least restraining the prevailing individualism.[22] Indeed, love for one another and the whole world constitutes a fundamental teaching and command of the Gospel of Christ.

There are numerous regions throughout the world, where for the general benefit of all, even when this contradicts the narrowly conceived interests of certain inhabitants, particular environmental prohibitions have been imposed. Therefore, by way of example, roads are forbidden from construction, while motor traffic is not permitted on the Greek island of Hydra and the Princes' Islands of Turkey. Moreover, *The Sayings of the Desert Fathers* mentions the example of a monk, who deliberately built his cell far away from the source of water in order to labor for his transportation of necessary resources.

Perhaps, then, it would not be far-fetched to demand of those who have voluntarily espoused the monastic life to be less concerned about their personal comfort and more concerned about the preservation of the natural beauty and the dimension of silence on Mt. Athos, the Holy Mountain. In any case, it is this aspect of silence that attracts those who renounce worldly cares. Therefore, the same persons are also obliged to preserve the cultural and natural environment of Mt. Athos by sacrificing certain material and mundane benefits for the sake of the more significant and spiritual benefits that are to come.

---

21. This brief comment is from an interview in the Greek journal *Forum* (January 2000).

22. Ecumenical Patriarch Bartholomew provided the message in this section for the opening of the Seminar on the Natural Environment of Mt. Athos (September 29, 1997).

As observed by Abba Isaac the Syrian, who is so beloved among the monks of Mt. Athos: No one ascends to heaven by means of comfort. Indeed, to extend our paternal recollection of this same Abba, we might remind you how he also says that God and the angels rejoice in heaven at the sight of ascetic labor, while the devil and his cohorts rejoice in hell at the sight of bodily comfort. However, Abba Isaac adds that every sacrifice for God leads to and results in everlasting gladness.

The Priest and the Scientist

The priesthood does not have an exclusive responsibility for the theological aspect of environmental concern; nor should science and technology feel self-sufficient in their scientific analysis of the subject.[23] What is purely scientific data should relate also to general theological information—both are necessary for a proper evaluation and appreciation of the ecological crisis. However, the opposite holds true as well: Theologians are called to cultivate a more comprehensive picture of scientific principles and demands in environmental issues. Indeed, we are convinced that such a mutual and common examination and exploration of theological and scientific methodology will result in the necessary essence and basis for particular recommendations and a more successful cooperation on the subject.

As we have stressed elsewhere, the Orthodox Church has always—based on the fourth and fifth century thought of the holy and God-bearing Fathers and Teachers of the undivided Church—emphasized cosmology together with, and never separated from anthropology. The classical thinkers of the Orthodox Church, such as the Cappadocian Fathers, never ignored the fundamentally eucharistic dimension of the creation of the world, which is returned to God in an act of thanksgiving and glorification. Material creation is good, both beautiful and blessed, expecting its eschatological transformation and the recapitulation of all in Christ, the eternal Word.

---

23. This section contains the text of a message from Ecumenical Patriarch Bartholomew for the conference "Orthodoxy and the Environment," held in Kavala, Greece (September 7, 1993).

POPULATION AND ENVIRONMENT

The most holy and Patriarchal Ecumenical Throne . . . expresses its satisfaction inasmuch as the Society of Population Studies [in Greece] has opportunely related the environmental crisis with the crucial and difficult subject of demographic explosion throughout the world.[24]

Indeed, in accordance with the opinion of specialist environmentalists and intellectuals, the breathtaking increase in our planet's population will create impasses for future generations, given that, in the wise words of the Swiss philosopher Denis de Rougemont, "a cancer cannot be larger than the body that it possesses."

FAITH AND SCIENCE

Orthodox theology has traditionally avoided, at least as much as possible, the conflict between faith and science on account of the double gnosiological methodology of the Church Fathers, whose thinking was based on the ontological distinction between Creator and creation.[25]

Therefore, the Church rejoices in the scientific success of humanity. It stands with a sense of profound awe, attention, and prudence before the various means of research and application presented by contemporary genetics and mechanics, biomedicine, and biotechnology. It seeks to establish an essential dialogue with these and sees a common responsibility that it shares with these for the healing of a suffering and burdened world.

Nevertheless, it also reminds the world that whatsoever abolishes a sense of respect before creation as a miracle of divine love and freedom also distorts the biblical and patristic criterion of moral truths, disturbs human relations, restricts human freedom, and undermines personal uniqueness.

---

24. Ecumenical Patriarch Bartholomew offered this message for the conference in Thessalonika on the subject of population development in relation to the environmental crisis (June 4, 1995).

25. The message in this section was offered to the delegates of the international conference, organized in Constantinople by the Ecumenical Patriarchate and entitled "The Creation of the World and the Creation of Humanity: challenges and problems before the 21st century," (September 1, 2002).

Scientific achievements, which soothe human pain, promote human health, extend human life, and generally improve human conditions, truly constitute divine gifts for the world. However, they are not ends in themselves. Nor are they excuses for an arrogant transcendence that reduces the soteriological perspective of the human person. The role of the Church is not to eliminate scientific progress but to realize human salvation and the promotion of moral values.

The expected life in the age to come is life in the kingdom of God. It is a life that begins "in time" and is fulfilled "in eternity." Historical time becomes a place of encounter with eternal reality. The salvific "now" of the kingdom is found in the Church, while the "always" of the Church exists in the heavenly kingdom. This life is not merely a preparation for the expected kingdom, but actually shares in this reality through faith in the kingdom, through the sacraments of the Church, and through the commandments of God.

Saints and the World

Our Church supports an attitude of thanksgiving toward creation, rather than an attitude of egoism that abuses the natural resources and life of the world.[26] It condemns greed, avarice, limitless acquisition, and uncritical consumerism, which sometimes reaches the degree of hybrid arrogance and insanity.

In this respect, we have the luminous examples of the saints, who respected life and humanity, who befriended the animals and the birds, who positively influenced their environment and community, and who lived with simplicity and self-sufficiency.

Therefore, we call upon every officer and minister, and invite every clergyman and educator, irrespective of degree and position, as well as every artists and journalist, to work together in withstanding the captivity of over-consumption: "It is for freedom that Christ has set us free. Stand firm, then, and no longer submit to any yoke of slavery" (Gal. 5.1).

---

26. Ecumenical Patriarch Bartholomew provided this message on the occasion of Earth Day, June 1997.

Theology and Life

The knowledge of the truth about creation, as well as of the purpose of the world and humanity, contributes to the correct response toward the ecological problem.[27] The Church reveals the truth, and in this way contributes toward the solution of every problem, including the environmental one.

The problems of humanity derive from the inner world of the human person, which defines human behavior. As a Christian hierarch, as a person that does not exercise worldly power, we propose the perfect way of human holiness as the solution of all human problems. It is clear that not everyone will follow this way, and so it is also clear that the problems will continue to exist. The powerful and aggressively greedy person creates the poor person, in the same way as the lazy person also creates his own poverty.

There are no purely theological matters that do not also have practical implications for life. The content of each person's faith creates the psychological presuppositions that define his conduct. Theology, then, and life are intertwined. Even the alienation of life from theology is a theological act, albeit negative. We pray that everyone will come to know the truth that sets us free.

Spiritual Roots of a Scientific Issue

Behind the ecological problem, just as behind many other contemporary issues, there lies concealed a theological stance and attitude.[28] The alienation of the humanity in Western society from God, neighbor, and environment, as well as the emphasis on individualism and utilitarianism have led to the abuse of sacred creation and to the modern ecological impasse. Unfortunately, humanity has lost the eucharistic relationship between the Creator God and the creation; instead of a priest and king, humanity has been reduced to a tyrant and abuser of nature.

---

27. These comments were provided as part of an interview with Anders Laugesen (January 2, 2001).

28. The text here served as the foreword to "Orthodoxy and the Natural Environment: a university symposium" (Thessalonika, 1997) 19.

Today, at the initiative of the Ecumenical Patriarchate, we observe a tendency for people to review their relationship with the environment and their use of the natural resources. Numerous critical studies relating to the erroneous ways of development are seeing the light of publicity. Many efforts are being taken for the cultivation of a technology that is respectful and rational toward creation. And we hope that all of these include indications of the concern of contemporary people to not only save the threatened natural environment but also to change their attitude toward God and the created world.

COLLECTIVE RESPONSIBILITY

TAGBLADET: In your opinion, Your All Holiness, how significant are the ecological issues for contemporary society?[29]

HIS ALL HOLINESS ECUMENICAL PATRIARCH BARTHOLOMEW: The Orthodox Church has always been conscious of the unity of humanity with the natural environment. Such awareness derives from the universal concept of the world and the inviolable interdependence of the parts with the whole, and vice versa. The Ecumenical Patriarchate has long ago said, and often continues to repeat, that God entrusted to humankind the supreme mission of celebrating in a "priestly" manner within sacred creation. The disruption of this relationship that we observe in the last years but especially in the most recent years, as well as the burdening of the natural environment, explain and justify the increased significance that we attribute to ecological-related issues. For, these are not simply local and regional but in fact concern the entire planet and its immediate environs, threatening the regular function of our natural environment as well as the very survival of humanity and the other living organisms.

TAGBLADET: Which is the most significant ecological challenge today?

H.A.H.: We do not have—nor do we claim to have—the qualifications to determine the most important ecological issues. This is a task that belongs to the specialists, whose knowledge and opinion we should all seek, heed, and respect. The Ecumenical Patriarchate systematically and consistently seeks to achieve this by organizing—among other things—interdisciplinary scientific dialogues on an international level in order to promote mutual information,

---

29. Ecumenical Patriarch Bartholomew provided this interview to the Norwegian newspaper *Tagbladet* (April 2002).

an unprejudiced concentration on real issues, and the recommendation of appropriate solutions wherever this is possible.

What people appear to be increasingly conscious of—or else they are obliged to become conscious of—is the fact that the environmental problem is an ethical issue. Usually, it is the more visible issues that are brought to people's attention. Nevertheless, it is wrong to ignore the spiritual, ethical, or deontological issues. For, the ecological problem is primarily a matter of each person's attitude or conduct before that part of sacred creation, which has been divinely appointed as the dwelling-place of humanity. Attitude and conduct are, of course, first of all individual matters. However, they are also collective matters that concern the context of life and action as local communities, nations, international coalitions, and even the global community define these.

The respect or violation of such contexts is a matter of moral order. From our own ecclesiastical perspective, we wish to highlight this essential dimension of the entire issue, as well as the urgent need to confront any indifference, wherever this may appear. For, indifference entails inaction, which in turn encourages further abuse in dealing with the causes that originally provoke and preserve this indifference.

TAGBLADET: In what way is the Church able to contribute, whether on the local or the international level?

H.A.H.: From what we have stated above, a partial response has already been given to your question. We believe that the responsibilities are many, just as the potential of the local ecclesiastical and in general the religious communities is great. Within our own Orthodox Church, each time we celebrate the Eucharistic Divine Liturgy, we chant with thanksgiving to God, saying: "Thine own [gifts] of Thine own, we offer to Thee." That is to say, we confess that the elements of Holy Communion, the bread and the wine, which are offered by the faithful for sanctification, recapitulate the entire creation that belongs to God and is offered to Him in thanksgiving. When these elements are transformed through the Holy Spirit into the Body and Blood of Christ, they are distributed to those who participate unto forgiveness of sins and life eternal. What is of special and creative significance is the spiritual exercise among the faithful of introspection and intense understanding of the unity between the mystery of creation, the mystery of life and the mystery of salvation. We feel that it is only from the depth of such a spiritual contemplation that humanity is able to behold the beauty of nature, to become sensitive to the harmony therein, and to wonder humbly and obediently at the sacred laws, which govern the function of the microcosm and the macrocosm. This humility, which ensues from a sense of wonder and awe, always inspired poets, hymnographers, and artists. It is also perhaps the only source of power that can enable humanity to confront its egocentrism, ugliness, greed, and arrogance.

TAGBLADET: Did the award of the Sophie Prize surprise you?

H.A.H.: We were gladly surprised. For, in our time, people do not usually place great hope in the Church, especially concerning issues that are considered "secular." This is why we regard the award of this prize as constituting a partial correction of this image. Moreover, our joy upon hearing this gracious decision offers us an encouragement and renewed starting point for further concern for the protection of our natural environment.

TAGBLADET: Is this your first visit to Oslo? Moreover, how does Your All Holiness intend to use the prize money?

H.A.H.: We visited Norway several years ago on the occasion of the Faith and Order Commission of the World Council of Churches and later returned there to participate in the events organized on the occasion of the millennial anniversary since the proclamation of the Christian Gospel in that country. We shall distribute the prize money to the needy children of Africa through Unicef, to the "street children" in Istanbul and Athens, as well as for the support of ecological initiatives.

ORTHODOXY AND ENVIRONMENT

It may surprise you that the first among hierarchs of the Orthodox Church is profoundly concerned with environmental issues.[30] The reason for this is that the Orthodox Church believes that the creation of God, both natural and spiritual, is "very good" (see Gen. 1.31) and that humanity is obliged to cultivate and to preserve this beautiful world, within which God placed us as rulers and providers, but not as unreasonable and abusive rulers.

Naturally, we distinguish the position of humanity from that of other created beings. We do not equate human beings with the other members of our ecosystem. This is why we regard humanity as being responsible for the rational regulation of the ecosystem and invite our fellow human beings to assume the necessary measures recommended by science for the improvement of the environment.

---

30. This section contains excerpts from a press conference with a number of journalists in Oslo, Norway, prior to receiving the Sophie Prize (June 12, 2002).

The Orthodox Church does not regard the material creation as evil (as the Gnostics did), nor the body as the prison of the soul (as Plato did). It believes that humanity in its entirety, body and soul alike, is destined for eternity and, is therefore sanctified, and resurrected with the body, which is transfigured on the one hand into a different form, yet which remains a body on the other hand.

The Orthodox Church is also concerned for the material needs of the world, exercising charity and healing illnesses (just as Christ did), caring for those who are hungry and naked, praying for seasonable weather and an abundance of the fruits of the earth. It is a Church that embraces the whole of creation.

In its sensitivity, the Orthodox Church feels compassion also for the inanimate nature and is unable to tolerate any harm that comes upon it. Therefore, within the context of recognizing nature as God's creation and humanity as nature's sustainer, with particular limitations that safeguard the abuse of creation, we have assumed the responsibility of raising people's awareness on environmental issues.

There is no need to emphasize the global consequences of local actions for the environment; this is familiar to everyone. There is also no need to underline that, in our environmental endeavors, we do not seek to worship it but to serve humanity; this is clear to everyone. We simply wish to stress that it is our responsibility—according to the commandment of God and the voice of our conscience—to face the environmental crisis with discernment, knowledge, love, and sacrifice. We are losing time; and the longer we wait the more difficult and irreparable the damage.

CONCERN FOR THE ENVIRONMENT

IRISH TIMES: Your All Holiness, why are you concerned with the environment?[31]
HIS ALL HOLINESS ECUMENICAL PATRIARCH BARTHOLOMEW: There are those that believe that the Christian Church only cares for life beyond death. The actual truth is that, though of course she does primarily devote herself to man and to his eternal salvation, she is also concerned with

---

31. The *Irish Times* conducted this interview with Ecumenical Patriarch Bartholomew on January 28, 2005.

man's life on this earth and with all of creation. Here is why: The Church cherishes all of creation because God put man in terrestrial Paradise "to dress it and to keep it" (Gen. 2.15), deriving the necessities for life from his "dressing" nature, thus causing her to produce the goods that are useful to him. Therefore nature does not remain entirely pristine, unaltered by man's activities, nor does she have equal value with man, as some would maintain. She is an instrument for his use, and a property for his usufruct.

But in respect of this use, God has posed certain limits for man that he ought not overstep. The limits are implicit in the "keeping," which means "not ravaging." Therefore, man's "dressing" of creation must not be destructive, but must revitalize and assist nature. Man must help nature to produce, and to renew herself: In other words he must preserve unharmed her own productive and revitalizing forces.

This duty remained for many centuries beyond the scope of careful study because the harmful effects of human intervention upon nature had not become widespread. But in recent times, as modern technology, and heavy industry with its pollutants, brought about major natural disasters and extensive environmental pollution, people of uncommon sensibility and perspicacity felt the need to react and to draw everybody's attention to the dangers that imperil humanity.

The Ecumenical Patriarchate and we personally support and sustain the effort to cause governments, industry, and ordinary people to become aware of the need to preserve the natural environment and the ecosystems of our planet, so as to keep it alive and in good health, that it may continue to yield its fruit in a sustainable manner. To this end, impelled by love of humanity rather than as under bondage to nature, we do our utmost, as far as it is possible, to engage in relevant actions. With the past presidents of the European Commission we have jointly held a number of environmental symposia in the Aegean and the Black Sea, on the Danube, in the Adriatic and the Baltic Sea. We have held several summer seminars in Halki on environmental issues. We have participated in a great number of environmental conferences all over the world. We have designated September 1st of each year as a day of prayer for the environment. And we have often spoken and written of our duty to "keep" the environment.

The fruit of all these efforts may not be measured, for there are indeed many other sensible people engaged in such work. We believe that these common efforts engaged in by all that are concerned over the environment, be they people, organizations, or governments, will improve the situation and avert major environmental disasters, and make a contribution toward restoring to health certain areas of the planet that have been damaged or are at risk.

That is what we hope for with all our heart so that, according to the wishes of God who is good, man's life on earth may be better. May it be so.

THE ROLE OF CHRISTIANITY

ULVHILD FAUSKANGER: During your visit in Norway, ecology will be in focus.[32] This is an area in which Your All Holiness has shown great engagement. Can you say something about the theological background for ecological concerns?

HIS ALL HOLINESS ECUMENICAL PATRIARCH BARTHOLOMEW: The environmental initiatives and concern of the Ecumenical Patriarchate date back to the mid-1980s and have focused around prayer for and protection of the environment, especially since the first Patriarchal encyclical was issued by our predecessor, the late Patriarch Demetrios, in September 1989, calling Orthodox Christians to celebrate and commemorate September 1st of each year as a day dedicated to supplications for the preservation of the natural environment. Five educational seminars were held on the island of Halki during the 1990s, while major international and interdisciplinary symposia have concentrated on the environmental problems of our seas. It is our conviction that—while the contemporary ecological issues faced by our world are undoubtedly connected to issues of social injustice, poverty, and war—they are primarily a result of a worldview and vision that have deviated from an understanding of this world as God's creation out of love.

U.F.: What is, in your view, the role of Christianity, and especially of the Orthodox Church, in regard to environmental questions?

H.A.H.: The role of the Christianity is to remind people that God created the world out of love in order that the entire world, with the human person as its center, may share in the divine beauty and sacredness as well as in order that the entire world, with the human person as its priest, may be referred back to God in an act of thanksgiving and praise. Christianity should also remind people that God "assumed flesh and dwelt among us" (Jn 1.14), appropriating our body and the body of the world. This "divine exchange," as the Church Fathers referred to it, is the way of the Eucharist, whereby bread and wine are offered to God, who then returns these as His own Body and Blood "for the

---

32. This interview with Ulvhild Fauskanger addressed the initiatives of the Ecumenical Patriarchate on the occasion of the North Sea symposium in June 2003.

life of the world" (Jn 6.51). The Orthodox Church retains a balanced eucharistic view, by claiming that this world is deeply imbued by God and by proclaiming that God is intimately involved with the world.

U.F.: What concrete results do you expect from the conference?

H.A.H.: The most tangible results that one may expect from such a meeting as this—apart from a deepened awareness of the particular environmental issues of the region as well as an increased knowledge of the ways in which science is able to investigate ecological problems and inform us about necessary responses that will benefit our environment—are a greater communication and cooperation between the various regional Churches and authorities. It is crucial for us to realize that no single person, discipline organization, institution, confession, or even religion can be blamed for or assume responsibility for resolving the crisis that we all face for the future of our children and the survival of our planet.

U.F.: Your All Holiness has also played a prominent role in the ecumenical movement. What are, in your view, the most important contributions of the Orthodox Church to this movement?

H.A.H.: The Ecumenical Patriarchate has—from its very own nature and from the very outset of the ecumenical movement—been directly and actively involved in participating in and promoting ecumenical relations between the various confessions (and even among the diverse religions). Well before the establishment of the World Council of Churches, the Ecumenical Patriarchate issued an encyclical letter "to the churches of Christ everywhere," recommending a "fellowship of churches" that would foster relations and rapprochement among them through study and cooperation. The Orthodox Church perceives the necessary response to the Lord's command and prayer "that they all may be one . . . so that the world may believe" (Jn 17.21), although it also recognizes the problems that beset us as well as the reality that only the Holy Spirit may reconcile us "in one body through the cross" (Eph. 2.16). The unique contributions of the Orthodox Church lie in its insistence that the One Church is indivisible and that denominationalism contradicts the unity that we seek. Moreover, the particular gifts brought to the ecumenical table by the Orthodox Church include the witness of the ancient and apostolic tradition, as well as its profound liturgy and spirituality.

U.F.: What do you see as the greatest challenges to the ecumenical movement today?

H.A.H.: In our modest opinion, the challenges that are faced by the churches within the ecumenical movement today are twofold. First, there is a need to reclaim the crucial and critical interconnection between doctrine and action, between faith and order, between theology and ethos. What we believe determines how we behave. The way that we pray is reflected upon the way that we treat our neighbor and our environment. There can be no separation between the image that we hold of our God, our world, or ourselves and the impact that we have upon other people and the natural environment. Second, churches are called to remember the essence and source of their existence and teaching. Unfortunately, so many confessional and denominational families now contain within themselves far more serious divisions than those that once divided them from one another. This surely places doubt upon their claims that historic divisions are maintained solely or primarily for the sake of truth. Therefore, we are invited to recall the original and essential purpose of the Church's being and life within a divided and disturbed world.

THE SOPHIE FOUNDATION

SOPHIE FOUNDATION: Your All Holiness, what is your opinion on climate change?[33]

HIS ALL HOLINESS ECUMENICAL PATRIARCH BARTHOLOMEW: Climate change affects everyone. While the data may be variously debated, the situation is clearly unsettling. Dramatic increases of greenhouse gases in our atmosphere—largely the result of fossil fuel burning—are causing global warming and in turn leading to melting ice caps, rising sea levels, the spread of disease, drought, and famine. The European heat wave of 2003 could be unusually cool by 2060, while the 150,000 people that the World Health Organization conservatively estimates are already dying annually because of climate change will be but a fraction of the actual number.

It is painfully evident that our response to the scientific testimony has been generally reluctant and gravely inadequate. Unless we take radical and immediate measures to reduce emissions stemming from unsustainable—in fact unjustifiable, if not simply unjust—excesses in the demands of our lifestyle, the impact will be both alarming and imminent.

Religious leaders throughout the world recognize that climate change is much more than an issue of environmental preservation. Insofar as human-induced, it is a profoundly moral and spiritual problem. To persist in the

---

33. Submitted by His All Holiness as a former recipient of the Sophie Prize, in response to questions addressed by the Sophie Foundation in January 2007.

current path of ecological destruction is not only folly. It is no less than suicidal, jeopardizing the diversity of the very earth that we inhabit, enjoy, and
share. We have repeatedly denounced it as a sin against God and creation.

Ecological degradation also constitutes a matter of social and economic
justice, for those who will most directly and severely be affected by climate
change will be the poorer and more vulnerable nations (what Christian Scriptures refer to as our "neighbor") as well as the younger and future generations
(the world of our children, and of our children's children).

S.F.: From what we know, what are we now called to do?

H.A.H.: There is a close link between the economy of the poor and the ecology
of the planet. Conservation and compassion are intimately connected. The
web of life is a sacred gift of God—ever so precious and ever so delicate.
We must serve our neighbor and preserve our world with both humility and
generosity, in a perspective of frugality and solidarity alike. After the great
flood, God pledged never again to destroy the world: "As long as the earth
endures, seedtime and harvest, cold and heat, summer and winter, day and
night, shall not cease" (Gen. 8.22). How tragic, however, it would be if we
were the ones responsible for their destruction. The footprint that we leave
on our world must be lighter, much lighter.

Faith communities must undoubtedly put their own houses in order; their
adherents must embrace the urgency of the issue. This process has already
begun, although it must be intensified. Religions realize the primacy of the
need for a change deep within people's hearts. They are also emphasizing the
connection between spiritual commitment and moral ecological practice.
Faith communities are well placed to take a long-term view of the world as
God's creation. In theological jargon, that is called "eschatology." Moreover,
we have been taught that we are judged on the choices we make. Our virtue
can never be assessed in isolation from others, but is always measured in solidarity with the most vulnerable. Yet churches, mosques, synagogues, temples,
and other houses of worship consume a fraction of energy compared to manufacturing industries, modern technologies, and commercial companies.

Breaking the vicious circle of economic stagnation and ecological degradation is a choice with which we are uniquely endowed at this crucial moment
in the history of our planet. The destructive consequences of indifference and
inaction are ever more apparent. At the same time, the constructive solutions
to mitigate global warming are increasingly merging. Government, businesses,
and religious institutions pursue cooperation, converging in commitment and
compelling people to act. Only together can a problem of this magnitude be
addressed and resolved. The responsibility as well as the response is collective.

S.F.: What are the urgent global challenges facing our world at this time?

H.A.H.: The urgent global challenges that we face in the future are summed up in the term "reconciliation." Humanity is called to build bridges and work toward peaceful coexistence in order to sustain the very survival of our planet. This obligation is, in fact, of unique significance inasmuch as humanity finds itself before a historical crossroads: For the first time in the history of the world, humanity has the privilege of choosing how to direct the future of creation. What is our divine gift and human birthright, namely the protection and preservation of humanity and the natural environment, is now a responsibility of enormous implications and consequences. How will we choose to respond?

First of all, it is clear that the natural environment is crying for liberation from the abuse and exploitation to which it has been exposed over the centuries and especially in recent decades. This is why, since 1995, we have also held six international, interdisciplinary, and interfaith symposia, focused on particular water bodies of the planet, in order to address specific environmental problems of the region and their implications for the global situation that we encounter. Thus, we have gathered political leaders, religious representatives, and journalists in order to explore ways of protecting creation and promoting an awakening of consciousness with regard to environmental issues. The truth is that, unless we change our perception of the world—in order no longer to exploit and abuse material and natural resources as if these were limitless—then future generations will no longer be able to enjoy the same gift of beauty that we have received from God. It is our firm conviction that not only is this our responsibility toward future generations but it is also our obligation to God and creation itself. It is not too late to act. But we must do so with a sense of urgency.

Moreover, we must do so together; for, our response to the ecological crisis is of common concern to people of all races, faiths, and disciplines. This leads us to a second global challenge of urgent importance, namely the peaceful coexistence of all peoples, irrespective of race, gender, and creed. It is our firm conviction that it is not religious differences that create conflicts among human beings. After all, if indeed human conflict was caused by the differences among religions, then there would be no tension or conflict among the faithful of the same religion. Yet, we know plenty of conflicts and wars even among faithful of the same religion, even within the same region. In our times of tension and conflict, many Westerners and other secularists single out religion as the forum to blame as the problem. As global politics intensify, they argue against religion and to some degree they may be right. But religion is not the point; religion is not the source; religion is not cause of the growing tensions throughout the world. As we have so often repeated, a war waged in the name of religion is a war waged against religion.

Therefore, religious leaders are responsible before God to preserve the teachings and traditions of our faith, to whatever degree we are of course cognitive of this faith—which implies the conscious rejection of any projecting of personal tastes seeking to replace the will of God. Furthermore, we are obligated humbly to demonstrate a profound respect toward believers of all religions, allowing our fellow human beings to journey on their own personal path to God, as they understand the will of God without on their part interfering with the journey of anyone else. This kind of profound respect on the part of one person toward the religious journey and conviction of another is the foundational responsibility of each of us. It is also the fundamental presupposition of peaceful coexistence and good will among people.

When we embark honestly on this journey of reconciliation—with one another, with people of every race and religion—then we begin healing divisions and building bridges between God and creation, between Christians, Jews, and Muslims, as well as between the rich and poor within one and the same society. This means that the response to the most global challenge begins within the human heart and is ultimately possible only through the grace of God.

ECONOMIC VALUES AND ECOLOGICAL VIGILANCE

HARRI SAUKKOMAA: Your All Holiness, how do economic values relate to spiritual values?[34]

HIS ALL HOLINESS ECUMENICAL PATRIARCH BARTHOLOMEW: Economic values are not at the summit of the Gospel values. The Lord said: "Seek first the Kingdom of God and His righteousness, and all these will be added unto you," (Matt. 6.33). Contemporary man moves in the opposite direction: He has deified the economy; and the economy, the golden calf set up and worshipped in the past on Mt. Sinai by those journeying to the Promised Land, has betrayed him! Yet, Christ, who cares with so much interest,

---

34. This interview with journalist Harri Saukkomaa was conducted in Finland in March 2009.

even for the smallest birds in the sky and the ephemeral wildflowers, as well as for the humblest of his creatures, is always here, prepared to assume this failure of ours as well, in order to lend us a hand of support, so long as we do not give our heart to mammon but instead entrust ourselves to His paternal love and providence.

H.S.: Why should we care for the natural environment?

H.A.H.: We should respect God's creation because disrespect of creation is tantamount to disrespect to the Creator; it is hubris! And hubris is always followed by nemesis. The world is our home. If we fail to take care of it in a loving manner, if we fail to preserve it with love, if we fail to protect it from every danger or continue to undermine it, then it will fall and destroy us. After all, there are other generations that will follow us; they, too, must find the home upright and strong enough to protect them.

H.S.: Are you optimistic that people will heed your advice?

H.A.H.: Unfortunately, the image that is presented globally does not justify much enthusiasm. People are constantly plunging into materialistic vanity and hedonism. "The care of this world and the deceitfulness of riches" (Matt. 13.22) stifle the Word of God that is consequently rendered fruitless, while a universal sense of hedonism is promoted through a myriad of means, misleading great numbers and especially the younger generation that is easily distracted to a way of life with a destructive message: "Let us eat and drink; for tomorrow, we die" (1 Cor. 15.32). However, the saints teach us the priority of the spirit and of virtue, offering us the example of a life that is abstinent, simple, ascetical (through fasting, frugality, and prayer), that transcends self-love; they teach us of a love and practical care for our neighbor, of a life that constantly looks upward, which in the final analysis is man's basic characteristic. And beyond the saints, of course, Christ Himself teaches us the very same, so that we have before us both the example and the "way."

H.S.: In your opinion, what is the greatest sin that people face today?

H.A.H.: The greatest of sins is *akedia*, or sloth! Namely, an unhealthy indifference toward what matters in life.

THE CONTRIBUTION OF ORTHODOXY

TERTIO: Can you describe the specific contribution of Orthodoxy to Christianity in general?[35]

---

35. This section provides the text of an interview with *Tertio*, Christmas 2009.

HIS ALL HOLINESS ECUMENICAL PATRIARCH BARTHOLOMEW: The power of the heart. This is the unique contribution of Orthodox Christianity. It is the assurance that the heart contains all of heaven and earth; it embraces the whole human person. It includes the body, mind, and spirit; it involves the reason, the will, and the emotions. "Within the heart," says St. Macarius of Egypt, "there lie unfathomable depths." Eastern Christianity has always treasured the heart as the place of convergence and conversation for all living things and for the whole of created nature. When the heart is at peace, the human and natural environments are reconciled. When the heart is at prayer, divine energy invigorates all of creation.

TERTIO: Your All Holiness is well known as the "Green Patriarch." How can Christians contribute to save nature and fight against climate change?

H.A.H.: Although little agreement appears to have been secured at Copenhagen with the recent UN Summit, one thing is certain: People have understood that words alone will not heal our broken environment. What is crucial is the ability to practice what we preach, to put into application all the wonderful words that we proclaim about climate change and global warming. By learning to sacrifice (that is to say, by learning to surrender some of our own luxuries) and beginning to share (that is to say, beginning to appreciate what the world needs), we will take the first steps toward responding to the unprecedented ecological crisis. We have repeatedly stated in the past that the crisis that we face is not so much a crisis of nature as it is a crisis of our lifestyle, which requires critical evaluation and radical change. In theological language, it implies repentance for the way we have ignored the sinful impact of our wasteful pollution.

GREEN THEOLOGY AND THE UNITED STATES

ODYSSEY: Your All Holiness, you have been a champion of the natural environment, even proclaimed the "Green Patriarch."[36] How are we called to respect God's creation in our lives?

HIS ALL HOLINESS ECUMENICAL PATRIARCH BARTHOLOMEW: Climate change was one of the foremost issues that we raised in our recent conversation with President Obama. Sometimes, our faithful are scandalized that we elevate the ecological crisis to such a level. However, we do not regard this matter as separated from the rest of our sacramental worldview and spiritual way. The natural environment provides us with a broader, panoramic vision

---

36. This interview appeared in *Odyssey Magazine* (January 2010).

of the world. We believe that in general nature's beauty leads us to a more open view of the life and created world, somewhat resembling a wide-angle focus from a camera, which ultimately prevents us human beings from using or abusing its natural resources. It is through the spiritual lens of Orthodox theology that we can better appreciate the broader aspects of such global problems as the threat to ocean fisheries, the disappearance of wetlands, the damage to coral reefs, or the destruction of animal and plant life.

The primary purpose of our visit to the United States was the eighth international and interdisciplinary symposium on the Mississippi River, where—in light of the upcoming environmental summit in Copenhagen—we sought to raise awareness of the intimate connections between ecological destruction, social justice, and political action, for the way we relate to material things directly reflects the way we relate to God. The sensitivity with which we handle worldly things clearly mirrors the sacredness that we reserve for heavenly things. And this is not simply a matter that concerns us as individuals; it concerns us collectively as members of church and society.

CLIMATE CHANGE AND POLITICAL RESOLVE

POLITICA NEWS JOURNAL: Because of your long-standing commitment to the field of ecology, you have been called the "Green Patriarch."[37] Are you content with what has been agreed on at the conference on climate change in Copenhagen? Do you think the world leaders could have achieved more?

HIS ALL HOLINESS ECUMENICAL PATRIARCH BARTHOLOMEW: Over the past two decades of our ministry, we have come to appreciate that one of the most valuable lessons to be gained from the ecological crisis is neither the political implications nor the personal consequences. Rather, this crisis reminds us of the connections that we seem to have forgotten between previously unrelated areas of life. For, the environment unites us in ways that transcend religious and philosophical differences as well as political and cultural differences. Paradoxically, the more we harm the environment, the more the environment proves that we are all connected.

Of course, there were great expectations that meaningful progress might be made as a result of the United Nations Climate Change Conference that took place in Copenhagen this month. However, Copenhagen is only

---

37. Ecumenical Patriarch Bartholomew provided this interview to *Politica News Journal* in Serbia (December 2009).

the beginning of a long process—for individuals, communities, and nations worldwide. It is a crucial promise to care for the planet in ways we never imagined as necessary before. However, it is more than a matter of care or concern. As faithful Orthodox, heirs of the tradition of the Desert Fathers and Mothers, we would underline the importance of asceticism as sacrifice.

Sacrifices will have to be made by all. Unfortunately, people normally perceive sacrifice as loss or surrender. Yet, the root meaning of the word has less to do with "going without" and more to do with "making sacred." Just as pollution has profound spiritual connotations, related to the destruction of creation when disconnected from its Creator, so too sacrifice is the necessary corrective for reducing the world to a commodity to be exploited by our selfish appetites. When we sacrifice, we render the world sacred, recognizing it as a gift from above to be shared with all humanity. Sacrifice is ultimately an expression of gratitude (for what we enjoy) and humility (for what we must share).

P.N.J.: For all the numerous appeals, we still live in a world full of conflicts and injustice. Are the rich and the powerful ready to hear the "low voice" of the Gospel?

H.A.H.: In our efforts for the preservation of the natural environment, it is easy to overlook our care for the welfare of the poor. It is not only the "other rich people" that we must question; we must question ourselves and consider what we have to offer other people. To paraphrase the Lord's Parable of the Good Samaritan, if each one of us is not identified with the Good Samaritan, then we are behaving like the highway robber in the story. We must ask ourselves some difficult questions about our concern for other human beings and about our way of life and daily habits. What are we prepared to surrender in order to learn to share? When will we learn to say: "Enough!"? How can we direct our focus away from what we want to what the world and our neighbor need?

The truth is that we tend—somewhat conveniently—to forget situations of poverty and suffering. And yet, we must learn to open up our worldview; we must no longer remain trapped within our limited, restricted point of view; we must be susceptible to a fuller, global vision. Tragically, we appear to be caught up in selfish lifestyles that repeatedly ignore the constraints of nature, which are neither deniable nor negotiable. We must relearn the sense of connectedness. For we will ultimately be judged by the tenderness with which we respond to human beings and to nature.

RELIGION AND SCIENCE

TRUD DAILY: You are often referred as the "Green Patriarch" because of your support given to environmental causes.[38] There are ongoing debates in Bulgaria and all over Europe related to the GMO.[39] For the time being the opinion that they should be banned prevails. What is your opinion on the issue?

HIS ALL HOLINESS ECUMENICAL PATRIARCH BARTHOLOMEW: In recent decades, we have understood very clearly that the way we worship and pray to God cannot be disassociated from the way we lead our lives and treat our planet. We are called to protect the natural resources of our world and not exploit or abuse the creation of God. In this respect, whereas the two disciplines have historically been suspicious of and even hostile to one another, religion has been encouraged—almost obliged—to dialogue with science. Therefore, we feel that the key word in your question is "debate." It is mandatory for religious and scientific representatives to be in continuous dialogue in order gradually to converge on the issues that are critical for our time, which includes the question of genetically modified organisms. This is precisely why, since 1995, the Ecumenical Patriarchate has inspired and initiated the Religious and Scientific Committee, which has in turn to date organized eight international, interfaith, and interdisciplinary sea-borne symposia on the Aegean and the Black Seas, the Danube River and the Adriatic Sea, the Baltic Sea and the Amazon River, the Arctic, and the Mississippi River. We are all facing the environmental crisis together. It is only together that we shall respond to and resolve it.

ENVIRONMENTAL AWARENESS

JULIAN CHRYSSAVGIS: Your All Holiness, what are you general views on the Climate Change Conference in Copenhagen next December?[40]

HIS ALL HOLINESS ECUMENICAL PATRIARCH BARTHOLOMEW: Our emphasis has always been that ecological issues do not merely affect one group of people or one nation or only "developed" countries or just the West. We are all in this together. Moreover, the solution will come from all of us and affect all of us in one way or another. This is why our ecological symposia

---

38. Ecumenical Patriarch Bartholomew spoke of religion and science in this interview with *Trud Daily* in Bulgaria (July 2010).

39. Genetically modified organisms.

40. The following interview with Julian Chryssavgis appeared in a longer article, which was published in *Today's Zaman* (August 2009).

are always international, interreligious, and interdisciplinary. The ecological crisis is not merely an economic or a political problem. Yet, it will take all of us—including those who formulate policies and define laws—to resolve it. This is why the contribution of Copenhagen is critical inasmuch as we pray that all of our political leaders—without exception—will embrace environmental protection once and for all.

J.C.: What would you like the readers of *Today's Zaman* to know about the forthcoming symposium in New Orleans?

H.A.H.: We were deeply moved when, in January of 2006, we first visited the destruction wrought by Hurricane Katrina in New Orleans and beyond. It was then that we were struck by how whatever we do in our individual lives—the choices we make as consumers of the earth's resources—has a direct impact on people and on the planet. We can no longer remain indifferent in our lifestyles of greed and selfishness. In depleting and defiling the earth, we are destroying life itself.

J.C.: Would you like to make any general remarks about your role or the role of Orthodoxy in raising environmental consciousness?

H.A.H.: Our concern for the environment does not result from any political or economic standpoint; nor again is it a superficial or sentimental romanticism. It arises from our effort to honor and dignify God's creation. We can no longer ignore the constraints of nature. So our humble ministry is to raise people's awareness about the importance of leaving a lighter footprint on our planet for the sake of generations to come. This is precisely why we have always emphasized that the ecological crisis is essentially a spiritual problem, which will only be resolved with a change of heart and mindset.

BELIEF AND ECOLOGY

SGI QUARTERLY: What were your first experiences of nature as a young person growing up in the Orthodox faith?[41]

HIS ALL HOLINESS ECUMENICAL PATRIARCH BARTHOLOMEW: As a young child, accompanying the priest of our local village to remote chapels on our native island of Imvros in Turkey, the connection of the beautiful mountainside to the splendor of God was evident. The environment provides a panoramic vision of the world, like the wide-angle lens of a camera, which prevents us from exploiting its natural resources in a selfish way. The recent

---

41. Ecumenical Patriarch Bartholomew spoke of belief and ecology in this interview, which appeared in *SGI Quarterly*, July 2010: "Religion and Ecology."

ecological disaster off the Gulf Coast in Louisiana reveals the consequences of ignoring this cosmic worldview. However, to reach a point of maturity toward the natural environment, we must take the time to listen to the voice of creation. It is unfortunate that we lead our life without noticing the environmental concert that is playing out before our eyes and ears. In this orchestra, each minute detail plays a critical role. Nothing can be removed without the entire symphony being affected. No tree, animal, or fish can be removed without the entire picture being distorted, if not destroyed.

SGI: Why are the ecological issues facing the world to be considered moral issues?

H.A.H.: We have repeatedly stated that the crisis that we are facing in our world is not primarily ecological. It is a crisis concerning the way we perceive the world. We are treating our planet in a selfish, godless manner precisely because we fail to see it as a gift inherited from above; it is our obligation to receive, respect, and return this gift to future generations. Therefore, before we can effectively deal with problems of our environment, we must change the way we regard the world. Otherwise, we are simply dealing with symptoms, not with their causes. We require "a new heavenly" worldview if we are to desire "a new earth" (Rev. 21.1).

SGI: You have consistently gone out of your way, through organizing several international interfaith symposia, to bring scientists and people of different churches within Christianity as well as different religious faiths and practices together to discuss ecological issues and visit sites around the world. What is your motivation for engaging in dialogue and building bridges?

H.A.H.: We are deeply convinced that any appreciation of the environmental concerns of our times must occur in dialogue with other Christian confessions, with other religious faiths, as well as with scientific disciplines. This is why, in 1994, we established the Religious and Scientific Committee. For, just as we share the earth, so do we share the responsibility for our pollution of the earth and the obligation to find tangible ways of healing the natural environment. To date, the Religious and Scientific Committee has hosted eight international, interdisciplinary, and inter-religious symposia to reflect on the fate of the rivers and seas, as well as to force the pace of religious debate on the natural environment. These symposia have gathered leading scientists, environmentalists, and journalists together with senior policymakers and representatives of the world's main religious faiths in an effort to draw global attention to the plight of the Aegean Sea, the Black Sea, the Danube River, the Baltic Sea, the Adriatic Sea, the Amazon River, the Arctic Sea, and the Mississippi River. Participants meet in plenary, workshop, and briefing sessions, hearing a variety of speakers on various environmental and ethical themes. Delegates also visit key environmental sites in the particular region of the symposium.

SGI: What, in your view, is your greatest success in these efforts?

H.A.H.: The greatest success undoubtedly lies in the process itself of gathering leaders of various disciplines. In the past, scientists were ignorant of religion, while theologians were suspicious of scientists. It is a matter of sincere effort and humility to sit beside representatives of other faith communities and scientific disciplines in order together to embrace accountability for the pollution of our planet and assume responsibility for the future of our world. The Religion, Science, and Environment symposia have taught us that God's creation can provide a common interest and concern, a universal objective and aspiration. All of us would like to leave behind a better world for our children.

SGI: Are there specific concepts in Orthodox Christianity that can help us in changing our attitude in relation to the environment?

H.A.H.: We believe that, as God's creation, this world is a sacred *mystery*, which in and of itself precludes any arrogance of *mastery* by human beings. Indeed, exploitation of the world's resources is identified more with Adam's "original sin" than with God's wonderful gift. It is the result of selfishness and greed, which arise from alienation from God and an abandonment of the sacramental worldview. We have, therefore, first of all proposed that the concept of sin be expanded—beyond the merely psychological or social—to include every act of pollution and misuse of God's creation. In this regard, we would add two other concepts. A proper use of the earth's natural resources implies a spirituality of "gratitude" (or Eucharist) and an ethos of "self-restraint" (or asceticism), so that we no longer willfully consume every fruit, but instead manifest a sense of frugality from some things for the sake of valuing all things. Then, we shall learn to care for plants and animals, for trees and rivers, for mountains and seas, for all human beings and the world.

SGI: How can religion come to be seen as a "savior of a humanity" not a threat to humanity?

H.A.H.: On the sixth day of creation, God created man and woman in His divine image and likeness. Yet, what most people overlook is that the sixth day is not dedicated to the formation of Adam alone. That sixth day was shared with "living creatures of every kind; cattle and creeping things and wild animals of the earth" (Gen. 1.24). This close connection between humanity and the rest of creation is a powerful reminder of our intimate relationship with the environment. While there is undoubtedly something unique about our creation in God's image, there is more that unites us than separates us, not only as human beings but also with creation. It is a lesson we have learned the hard way in recent decades. The connection is not merely emotional; it is profoundly spiritual, providing a sense of continuity and community as well as an expression of identity and compassion with all of creation. In the seventh century, Abba Isaac the Syrian said: "A merciful heart burns with love for all creation: for human beings, birds, and beasts—for all God's creatures." When we recognize this connection, then we shall be instruments of peace and life, not tools of violence and death. Then, everything will assume its divine purpose, as God originally intended the world.

9

Declarations and Statements

RELIGIOUS DECLARATIONS AND STATEMENTS

Joint Declaration with Pope John Paul II, Symposium IV,
Venice-Vatican, June 10, 2002

IT IS NOT TOO LATE

We are gathered here today in the spirit of peace for the good of all human beings and for the care of creation.[1] At this moment in history, at the beginning of the third millennium, we are saddened to see the daily suffering of a great number of people from violence, starvation, poverty, and disease. We are also concerned about the negative consequences for humanity and for all creation resulting from the degradation of some basic natural resources such as water, air, and land, brought about by an economic and technological progress that does not recognize and take into account its limits.

Almighty God envisioned a world of beauty and harmony, and He created it, making every part an expression of His freedom, wisdom, and love (see Gen 1.1–25). At the center of the whole of creation, He placed us, human beings, with our inalienable human dignity. Although we share many features with the rest of the living beings, Almighty God went further with us and gave us an immortal soul, the source of self-awareness and freedom, endowments that make us in His image and likeness (see Gen. 1.26–31; 2.7). Marked with that resemblance, we have been placed

---

1. Common declaration on environmental ethics signed by Ecumenical Patriarch Bartholomew I in Venice during the closing ceremony at the Palazzo Ducale in Venice and co-signed by Pope John Paul II via satellite connection from his library in the Vatican.

by God in the world in order to cooperate with Him in realizing more and more fully the divine purpose for creation.

At the beginning of history, man and woman sinned by disobeying God and rejecting His design for creation. Among the results of this first sin was the destruction of the original harmony of creation. If we examine carefully the social and environmental crisis that the world community is facing, we must conclude that we are still betraying the mandate God has given us: to be stewards called to collaborate with God in watching over creation in holiness and wisdom.

God has not abandoned the world. It is His will that His design and our hope for it will be realized through our cooperation in restoring its original harmony. In our own time we are witnessing a growth of an *ecological awareness,* which needs to be encouraged so that it will lead to practical programs and initiatives. An awareness of the relationship between God and humankind brings a fuller sense of the importance of the relationship between human beings and the natural environment, which is God's creation and which God entrusted to us to guard with wisdom and love (see Gen. 1.28).

Respect for creation stems from respect for human life and dignity. It is on the basis of our recognition that the world is created by God that we can discern an objective moral order within which to articulate a code of environmental ethics. In this perspective, Christians and all other believers have a specific role to play in proclaiming moral values and in educating people in *ecological awareness,* which is none other than responsibility toward self, toward others, toward creation.

What is required is an act of repentance on our part and a renewed attempt to view ourselves, one another, and the world around us within the perspective of the divine design for creation. The problem is not simply economic and technological; it is moral and spiritual. A solution at the economic and technological level can be found only if we undergo, in the most radical way, an inner change of heart, which can lead to a change in lifestyle and of unsustainable patterns of consumption and production. A genuine *conversion* in Christ will enable us to change the way we think and act.

First, we must regain humility and recognize the limits of our powers, and most important, the limits of our knowledge and judgment. We have

been making decisions, taking actions, and assigning values that are leading us away from the world as it should be, away from the design of God
for creation, away from all that is essential for a healthy planet and a
healthy commonwealth of people. A new approach and a new culture are
needed, based on the centrality of the human person within creation and
inspired by environmentally ethical behavior stemming from our triple
relationship to God, to self, and to creation. Such an ethics fosters interdependence and stresses the principles of universal solidarity, social justice, and responsibility, in order to promote a true culture of life.

Second, we must frankly admit that humankind is entitled to something better than what we see around us. We and, much more, our children and future generations are entitled to a better world, a world free
from degradation, violence and bloodshed, a world of generosity and love.

Third, aware of the value of prayer, we must implore God the Creator
to enlighten people everywhere regarding the duty to respect and carefully
guard creation.

We therefore invite all men and women of good will to ponder the
importance of the following ethical goals:

To think of the world's children when we reflect on and evaluate our options
for action.

To be open to study the true values based on the natural laws that sustain
every human culture.

To use science and technology in a full and constructive way, while recognizing that the findings of science have always to be evaluated in the light of
the centrality of the human person, of the common good, and of the inner
purpose of creation. Science may help us to correct the mistakes of the past
in order to enhance the spiritual and material well-being of the present and
future generations. It is love for our children that will show us the path
that we must follow into the future.

To be humble regarding the idea of ownership and to be open to the demands
of solidarity. Our mortality and our weakness of judgment together warn
us not to take irreversible actions with what we choose to regard as our
property during our brief stay on this earth. We have not been entrusted
with unlimited power over creation; we are only stewards of the common
heritage.

To acknowledge the diversity of situations and responsibilities in the work for
a better world environment. We do not expect every person and every

institution to assume the same burden. Everyone has a part to play, but for the demands of justice and charity to be respected, the most affluent societies must carry the greater burden, and from them is demanded a sacrifice greater than can be offered by the poor. Religions, governments, and institutions are faced by many different situations, but on the basis of the principle of subsidiarity all of them can take on some tasks, some part of the shared effort.

To promote a peaceful approach to disagreement about how to live on this earth, about how to share it and use it, about what to change and what to leave unchanged. It is not our desire to evade controversy about the environment, for we trust in the capacity of human reason and the path of dialogue to reach agreement. We commit ourselves to respect the views of all who disagree with us, seeking solutions through open exchange, without resorting to oppression and domination.

It is not too late. God's world has incredible healing powers. Within a single generation, we could steer the earth toward our children's future. Let that generation start now, with God's help and blessing.

On Climate Change

GLOBAL WARMING AND GLOBAL ECONOMY

Although the data regarding climate change is sometimes debated, the seriousness of the situation is generally accepted.[2] Climate change affects everyone. Unless we take radical and immediate measures to reduce emissions stemming from unsustainable—in fact unjustifiable, if not simply unjust—excesses in the demands of our lifestyle, the impact will be both alarming and imminent.

Climate change is much more than an issue of environmental preservation. Insofar as human-induced, it is a profoundly moral and spiritual problem. To persist in the current path of ecological destruction is not only folly. It is no less than suicidal, jeopardizing the diversity of the very earth that we inhabit, enjoy, and share. Moreover, climate change constitutes a matter of social and economic justice. For those who will most directly and severely be affected by climate change will be the poorer

---

2. Geneva, 2004. Statement requested and adopted by the World Council of Churches.

and more vulnerable nations (what Christian Scriptures refer to as our "neighbor") as well as the younger and future generations (the world of our children, and of our children's children).

There is a close link between the economy of the poor and the warming of our planet. Conservation and compassion are intimately connected. The web of life is a sacred gift of God—ever so precious and ever so delicate. We must serve our neighbor and preserve our world with both humility and generosity, in a perspective of frugality and solidarity alike.

Faith communities must undoubtedly put their own houses in order; their adherents must embrace the urgency of the issue. This process has already begun, although it must be intensified. Religions realize the primacy of the need for a change deep within people's hearts. They are also emphasizing the connection between spiritual commitment and moral ecological practice. Faith communities are well placed to take a long-term view of the world as God's creation. In theological jargon, that is called "eschatology." Moreover, we have been taught that we are judged on the choices we make. Our virtue can never be assessed in isolation from others but is always measured in solidarity with the most vulnerable. Breaking the vicious circle of economic stagnation and ecological degradation is a choice with which we are uniquely endowed at this crucial moment in the history of our planet.

On Water

NATURE AND SPIRIT

Water is as fundamental in the natural life as it is in the spiritual world.[3] As the Book of Genesis says: "In the beginning, God created the heavens and the earth. Now the earth was a formless void, there was darkness over the deep, and God's spirit hovered over the water" (Gen. 1.1–2). Just as water is the essence of all life, water is also the primary element in the life of a Christian, where the sacrament of baptism marks the sacred source of the spiritual life.

The striking connection between the natural world and the world of the Spirit is indicated in the ceremony of the Great Blessing of the Waters,

---

3. Geneva, 2004. Statement requested and adopted by the World Council of Churches.

performed in the Orthodox Church on January 6, the Feast of Theophany, when we commemorate Christ's baptism in the Jordan River. The Great Blessing begins with a hymn of praise to God for the beauty and harmony of creation:

> Great are You, O Lord, and marvelous are Your works: no words suffice to sing the praise of Your wonders. . . . The sun sings Your praises; the moon glorifies You; the stars supplicate before You; the light obeys You; the deeps are afraid at Your presence; the fountains are Your servants; You have stretched out the heavens like a curtain; You have established the earth upon the waters; You have walled about the sea with sand; You have poured forth the air that living things may breathe. . . .

Water, then, signifies the depth of life and the calling to cosmic transfiguration. It can never be regarded or treated as private property or become the means and end of individual interest. Indifference toward the vitality of water constitutes both a blasphemy to God the Creator and as a crime against humanity. Through the pollution or contamination of the world's waters, the destruction is procured of the planet's entire ecosystem, which receives its life from unceasing communication, like communicating vessels, of the watery subterranean or supraterranean arteries of the earth.

PUBLIC MESSAGES

On World Oceans Day, Symposium V, Stockholm, June 2004

A COMMON HERITAGE

Distinguished participants in the fifth Religion, Science, and the Environment Symposium; assembled dignitaries; scholars; communicators; leaders of congregations; members of the press; and citizens of God's beautiful earth:

> Some of us are native speakers of Swedish, French, English, Finnish, Russian, or Greek, and some of languages even beyond Europe. Some of us are young, some middle-aged, and some have seen many seasons come and go. Some of us are representatives of academia, some of industry, some of

public interest organizations, NGOs, some of government and intergovernmental organizations. Some of us are Roman Catholics, some Protestants, some Jews, some Muslims, and some are Orthodox Christians.

But one thing we all share is a common heritage of humankind. We are united by water, which makes up 70 percent of our body and 70 percent of the earth's surface. All life depends on its nourishing power. Flowing water makes our planet unique among all the planets, as we know, in the universe. Water is a source of wonder and beauty, a cause of celebration and connection.

THE POWER OF WATER

Water cradles us from our birth, sustains us in life, and heals us in sickness. It delights us in play, enlivens our spirit, purifies our body, and refreshes our mind. We share the miracle of water with the entire community of life. Indeed, each of us is a microcosm of the oceans that sustains life. *Every person here, every person in our world, is in essence a miniature ocean.* The world's oceans, coastal waters, enclosed seas, and estuaries abound with the gift of life. The vast blue oceans are God's creation no less than the land, no less than these tiny oceans, and not less than we ourselves are.

And so it is that we have gathered together to remember the oceans on this eve of World Oceans Day. World Oceans Day celebrates the largest earthly realm of God's creation. Oceans were and are our medium of travel around the world. They link all peoples, coastal and landlocked, in study, in trade, in communication, in worship. The oceans provide one-sixth of the animal protein consumed by humans, more so than chicken, beef, mutton, or pork. Oceans generate nearly half of the oxygen we breathe and cleanse the atmosphere of much carbon dioxide that people, automobiles and power plants produce. Removing this carbon dioxide is vital because the human-caused increase of atmospheric carbon dioxide threatens our planet's biological diversity and our own human civilization.

The oceans are the earth's major shapers of climate, to whose existing patterns our societies have adapted. More than half of humanity lives within 100 kilometers of the coast, and where we live reflects our affinity for the gifts that the oceans provide. The health of the oceans is essential to human well-being, to biological diversity, and to the stability of our world's climate. And the oceans are home to countless species of life, from great whales that are loved by all to tiny creatures that only God knows. But the health of the oceans and seas is severely threatened. We overfish. We pollute. We have nearly exhausted our seas. The Baltic Sea, on which this city is built, suffers from and is being stifled by humans. Once we were few and the sea seemed vast; but now we are exceedingly numerous, and yet the Baltic has not grown. We surround it, and this surrounded sea feels the seasonal pulse of human life. Our species has harmed the life of this beautiful and once-bountiful sea.

THE PERIL OF THE OCEAN

The Baltic Sea is a microcosm of the world's oceans. It is one of many seas bordering the interconnected basins of the world's oceans. The Baltic's waters may have originated in rain fallen over Germany or Poland or Estonia, and brought to it by a hundred rivers. Nevertheless, when these waters leave the Baltic, they will most certainly touch the tropical shores of Indonesia, the ice-bound coasts of Antarctica, and the darkest waters of the Challenger Deep before the end of time. There is truly only one ocean, with all its parts interconnected, just as there is only one Spirit, which unites us all.

So what we do to the oceans, God's vast blue creation, we also do to God's other creations, including ourselves. Once we humans did not know that we could harm God's creation. The oceans, especially, seemed so vast as to be invulnerable. But now we know differently. We know how fragile is our precious earth and its oceans. We know how essential they are to sustaining our bodies, our minds, our hearts, and our spirits. And we know that the oceans, even in their vastness, are feeling the crushing burden of humans' callous ignorance. We fish their depths to exhaustion, we fill them with pollution, we reshape their shores with little concern for their ability to endure. From the Baltic Sea to the remotest Southern Ocean seamounts, we have reduced the oceans' miraculous biodiversity.

The oceans are in peril. They cannot protect themselves. But God has endowed humankind with the knowledge to rectify our mistakes, and we are, each one of us, given the choice of what we will do. To protect the oceans is to do God's work. To harm them, even if we are ignorant of the harm we cause, is to diminish His divine creation. We can stop overfishing and destructive fishing methods so that the miracle of the fishes will endure for future generations. We can stop pollution so that the seas can recover from poisoning and from life-choking nutrients produced by our cities and farms and industries. We can establish sanctuaries in the sea where we agree to do no harm of any kind.

If we can find the faith to love each other and to love God, then we can find the faith to help His vast water planet live and flourish. On this eve of World Oceans Day, we invite all of you to join us in pledging to protect the oceans as an act of devotion, whatever your religion may be. If we love God, we must love His creation.

On World Environment Day

"SERVING AND PRESERVING" CREATION

Today's World Environment Day is an opportunity as well as an invitation for all of us, irrespective of religious background, to consider the ecological crisis.[4]

In our time, more than ever before, there is an undeniable obligation for all to understand that environmental concern for our planet is not merely a romantic notion of the few. The ecological crisis, and particularly the reality of climate change, constitutes the greatest threat for every form of life in our world. Moreover, there is an immediate correlation between protection of the environment and every expression of economic and social life.

For our Orthodox Church, the protection of the environment as God's creation is the supreme responsibility of human beings, quite apart from any material or other financial benefits that it may bring. The almighty God bequeathed this "very beautiful" world (see Gen. 1.26) to humanity together with the commandment to "serve and preserve" it. Yet, the direct correlation of this divine mandate for the protection of creation to every aspect of contemporary economic and social life ultimately enhances the global effort to control the problem of climate change by effectively introducing the ecological dimension into every aspect of life.

With the opening of this third millennium, environmental issues—already evident since the twentieth century—acquired a new intensity, coming to the forefront of daily attention. According to the theological

---

4. June 5, 2005.

understanding of the Orthodox Christian Church, the natural environment is part of creation and is characterized by sacredness. This is why its abuse and destruction is a sacrilegious and sinful act, revealing prideful scorn toward the work of God the Creator. Humanity, too, is part of this creation. Our rational nature, as well as the capacity to choose between good and evil, bestow upon us certain privileges as well as clear responsibilities. Unfortunately, however, human history is filled with numerous examples of misuse of these privileges, where the use and preservation of natural resources has been transformed into irrational abuse and, often, complete destruction, leading occasionally to the downfall of great civilizations.

Indeed, the care for and protection of creation constitutes the responsibility of everyone on an individual and collective level. Naturally, the political authorities of each nation have a greater responsibility to evaluate the situation in order to propose actions, measures, and regulations that will convince our communities of what must be done and applied. Yet, the responsibility of each individual is also immense, both in one's personal and family life but also in one's role as an active citizen.

Thus, we call everyone to a more acute sense of vigilance for the preservation of nature and all creation, which God made in all His wisdom and love. And, from the See of the Ecumenical Patriarchate, we invoke God's blessing for World Environment Day, offering praise to the Creator of all, to whom is due all glory, honor, and worship.

On World Environment Day

THE IMPERATIVE OF FRUGALITY

Inasmuch as, at the Ecumenical Patriarchate, we have long been concerned about problems related to the preservation of the natural environment, we have ascertained that the fundamental cause of the abuse and destruction of the world's natural resources is greed and the constant tendency toward unrestrained wealth by citizens in so-called "developed" nations.[5]

The Holy Fathers of our Church have taught and lived the words of St. Paul, according to which "if we have food and clothing, we will be

---

5. June 5, 2010.

content with these" (1 Tim. 6.8), adhering at the same time to the prayer of Solomon: "Grant me neither wealth nor poverty, but simply provide for me what is necessary for sufficiency" (Prov. 24). Everything beyond this, as St. Basil the Great instructs, "borders on forbidden ostentation."

Our predecessor on the Throne of Constantinople, St. John Chrysostom, urges: "In all things, we should avoid greed and exceeding our need"[6] for "this ultimately trains us to become crude and inhumane,"[7] "no longer allowing people to be people, but instead transforming them into beasts and demons."[8]

Therefore, convinced that Orthodox Christianity implies discarding everything superfluous and that Orthodox Christians are "good stewards of the manifold grace of God" (1 Pet. 4.10), we conclude with a simple message from a classic story, from which everyone can reasonably deduce how uneducated, yet faithful and respectful people perceived the natural environment and how it should be retained pure and prosperous.

In the *Sayings of the Desert Fathers on the Sinai*, it is said about a monk known as the righteous George, that eight hungry Saracens once approached him for food, but he had nothing whatsoever to offer them because he survived solely on raw, wild capers, whose bitterness could kill even a camel. However, upon seeing them dying of extreme hunger, he said to one of them: "Take your bow and cross this mountain; there, you will find a herd of wild goats. Shoot one of them, whichever one you desire, but do not try to shoot another." The Saracen departed and, as the old man advised, shot and slaughtered one of the animals. But when he tried to shoot another, his bow immediately snapped. So he returned with the meat and related the story to his friends."

350. ORG

CHOICES AT A CRITICAL TIME

Breaking the vicious circle of ecological degradation is a choice with which we are uniquely endowed at this crucial moment in the history of

---

6. *Homily XXXVII on Genesis.*
7. *Homily LXXXIII on Matthew.*
8. *Homily XXXIX on 1 Corinthians.*

our planet.[9] We have traditionally regarded sin as being merely what people do to other people. Yet, for human beings to degrade the integrity of the earth by contributing to climate change, by stripping the earth of its natural forests or destroying its wetlands; for human beings to contaminate the earth's waters, land and air—these are sins. 350.ORG is repentance in action.

On the Declaration of Earth Rights

LISTENING TO THE VOICE OF CREATION

In recent years, we have learned—albeit painfully—that what we do to creation directly reflects how we pray to the Creator and how we relate to one another.[10] Over the last decades, we have come to understand—albeit reluctantly—that the sensitivity with which we handle worldly things clearly mirrors the sacredness that we reserve for heavenly things and the subtlety that is demanded in human conduct.

This is why, since the 1980s, the Ecumenical Patriarchate has sought to underline and champion the rights of the natural environment. As a result, therefore, we have continually and persistently declared that, unless we radically change our perception of the natural world, unless we reverse our self-centered estimation of the planet, then we cannot possibly appreciate the destructive impact of our exploitation and abuse.

Moreover, we have long declared that to commit a crime against the natural world is a sin. For human beings to cause species to become extinct or destroy the biological diversity of God's creation; to degrade the integrity of the earth by causing climate change, to strip the earth of its natural forests or destroy its wetlands; to contaminate the earth's waters, its land, its air, and its life—all of these are sins.

In order, however, to reach the point of maturity and confess that we have sinfully and inhumanely behaved toward the natural environment, we must take the time to listen to the voice of creation. And to do this, we must first be silent. Then, if we are honest with ourselves, we must

---

9. 350.org is a website and network, inspired by environmentalist and author Bill McKibben as a global movement to solve the climate crisis.

10. Istanbul, 2010.

search for ways to restrain our selfish desires and restrict our harmful attitude that has depleted and destroyed the natural resources of our planet.

Occasional Statements

Let Us Respond in Practice

OUR SURVIVAL AT STAKE

This week, political leaders converge on Copenhagen, Denmark in what is undoubtedly the most anticipated meeting on climate change in history.[11] Their meeting has already been preceded by a gathering of religious leaders and attended by numerous activists and concerned citizens.

Only last month, in light of the UN Summit in Copenhagen, the Ecumenical Patriarchate convened an international and interdisciplinary symposium in New Orleans on the Mississippi River to consider its fate within the global environment.

At first glance, it may appear strange for the leader of a religious institution concerned with "sacred" values to be so profoundly involved in "worldly" issues. After all, what does preserving the planet have to do with saving the soul? It is commonly assumed that global climate change and the exploitation of our nature's resources are matters that concern politicians, scientists, and technocrats. At best, perhaps, they are the preoccupation of interest groups or naturalists.

So the preoccupation of the Orthodox Church and, in particular, her highest administrative authority, the Ecumenical Patriarchate, with the environmental crisis will probably come as a surprise to many people. Yet, there are no two ways of looking at either the world or God. There can be no double vision or worldview. In our understanding, there can be no distinction between concern for human welfare and concern for ecological preservation.

Nature is a book, opened wide for all to read and to learn. It tells a unique story; it unfolds a profound mystery; it relates an extraordinary

---

11. General Statement issued at the Ecumenical Patriarchate in light of the United Nations Climate Change Conference in Copenhagen, December 7–18, 2009.

harmony and balance, which are interdependent and complementary. The way we relate to nature as creation directly reflects the way we relate to God as Creator. The sensitivity with which we handle the natural environment clearly mirrors the sacredness that we reserve for the divine. We must treat nature with the same awe and wonder that we reserve for human beings. And we do not need this insight in order to believe in God or to prove His existence. We need it to breathe; we need it for us simply to be.

At stake is not just our ability to live in a sustainable way, but our very survival. Scientists estimate that those most hurt by global warming in years to come will be those who can least afford it. Therefore, the ecological problem of pollution is invariably connected to the social problem of poverty, and so all ecological activity is ultimately measured and properly judged by its impact and effect upon the poor.

In our efforts, then, to contain global warming, we are admitting just how prepared we are to sacrifice some of our greedy lifestyles. When will we learn to say: "Enough!"? When will we direct our focus away from what we want to what the world needs? When will we understand how important it is to leave as light a footprint as possible on this planet for the sake of future generations? We must choose to care. Otherwise, we do not really care at all.

We are all in this together. Indeed, the natural environment unites us in ways that transcend doctrinal differences. We may differ in our conception of the planet's origin. But we all agree on the necessity to protect its natural resources, which are neither limitless nor negotiable.

It is not too late to respond—as a people and as a planet. We could steer the earth toward our children's future. Yet we can no longer afford to wait; we can no longer afford not to act. The decisions at Copenhagen must be a clear step forward. The world has clearly expressed its opinion; our political leaders must act accordingly. Deadlines can no longer be postponed; indecision and inaction are not options.

We are optimistic about the results at Copenhagen, quite simply because we are optimistic about humanity's potential. Let us not simply respond in principle; let us respond in practice. Let us listen to one another; let us work together; let us offer the earth an opportunity to heal and continue to nurture us.

The Moment Is Coming When It Will Be Too Late

NOT WHAT WE WANT, BUT WHAT THE PLANET NEEDS

Observing the leaders of our world as they gather in Copenhagen, we pray and hope that they realize just how late we have left it to restore our earthly home to health.[12]

There will come a point, and it may be very soon, when it is simply *too* late.

Our scientists talk of "tipping points" and "abrupt climate change." Our political leaders talk of the challenges that lie ahead. The Bible speaks of God's grace in giving us many, many chances. But it makes it clear the time will come for all of us when we have to face the consequences of our wrongdoing.

In the Gospel of St. Luke, the rich man in his fine robes, who ignored the beggar Lazarus at his gate, was condemned after death for his indifference and inaction. When he begged to be released from torment, he was told that it was too late.

According to the Gospel of St. Matthew, the questions asked of us all at the Last Judgment will not be about religious observances. We will be asked if we fed the hungry, gave drink to the thirsty, clothed the naked, and comforted the sick and the captives.

We must make the sacredness of life our priority.

It is life on earth, which is threatened, not just a certain way of living.

This is the very same earth that we are commanded not just to "till," but also to "preserve."

Our reckless consumption of resources—fuel, water, forests—threatens us with a climate Apocalypse. Burning more fuel than we need in a busy city, we may be contributing to a drought or flood in a place thousands of miles away. Scientists estimate that those most hurt by global warming in the years to come, are those who can least afford it.

In our understanding there can be no distinction between concern for human welfare and concern for ecological preservation.

To restore the planet we need a spirituality of humility and respect—one that leads us to inquire more deeply and to think of the impact of our actions on all of creation.

---

12. Statement to the "British Broadcasting Commission for the United Nations Climate Change Conference" in Copenhagen, December 7–18, 2009.

We have been privileged in recent years to lead our Religion, Science, and the Environment Symposia to meet many individuals whose lives are threatened by distant forces that they can neither understand nor control. We have stood on the banks of the Amazon and witnessed the destruction of the rainforest in the name of providing cheap food for the well-fed. We have stood and watched a great glacier of Greenland melt as its world was warmed by greenhouse gasses. A month ago in New Orleans we heard evidence of how hubris turned a natural event into a human catastrophe.

We must direct our focus away from what we want to what the planet needs. We must choose to care for creation; otherwise we do not really care at all.

Nature unites us, and while we may differ in our conception of the origins of our world, we all agree on the necessity to protect its future— our future. Let us offer the earth an opportunity to heal and continue to nurture us.

We can no longer afford to wait. Indecision and inaction are not options.

We are all living within the Mercy and Grace of God.

Our faith makes clear that we have a choice.

The time to choose is now.

On the Oil Spill on the Gulf of Mexico

FERVENT PRAYER AND SINCERE HOPE

Once again, in a matter only of a few years, the eyes of the world are turned with suspense toward the Gulf Coast.[13] Sadly, the oil spill is following a path similar to Hurricane Katrina and threatening the coast of Louisiana as well as neighboring states.

As citizens of God's creation, we perceive this monumental spill of crude oil in the oceans of our planet as a sign of how far we have moved from the purpose of God's creation.

Our immediate reaction is to pray fervently for the urgent and efficient response to the current crisis, to mourn painfully for the sacrifice of human life as well as for the loss of marine life and wildlife, and to support

---

13. Istanbul, April 2010. Entitled: "Sins against nature and God."

the people and communities of the region, whose livelihood directly depends on the fisheries of the Gulf.

But as the first bishop of the world's second-largest Christian Church, we also have a responsibility not only to pray, but also to declare that to mistreat the natural environment is to sin against humanity, against all living things, and against our creator God. All of us—individuals, institutions, and industries alike—bear responsibility; all of us are accountable for ignoring the global consequences of environmental exploitation. Katrina—we knew—was a natural calamity. This time—we know—it is a man-made disaster. One deepwater pipe will impact millions of lives in several states as well as countless businesses and industries.

Therefore, we must use every resource at our disposal to contain this disaster. But we must also use every resource to determine liability for the fact that 11 people have died and 5,000 barrels of oil are flowing daily into the delicate ecology of the Gulf of Mexico. In exchange for the benefits and wealth generated by deep underwater drilling, individuals, institutions, and industries assume responsibility for protecting the earth and its creatures from the well-known potential hazards. In this instance, they have clearly failed in those responsibilities; that failure must be acknowledged and strong measures taken to avert future catastrophes.

Although we are halfway around the world from this incident, our interest in it is deeply personal. We visited Louisiana and its bayous only four months after its devastation by Hurricane Katrina and we returned there just last October to convene our eighth Religion, Science, and the Environment Symposium, "Restoring Balance: The Great Mississippi," in New Orleans. At that time, we noted:

> Although the time we have been on the planet is insignificant in the context of the life of the planet itself, we have reached a defining moment in our story.
>
> Let us remember that, whoever we are, we all have our part to play, our sacred responsibility to the future. And let us remember that our responsibility grows alongside our privileges; we are more accountable the higher we stand on the scale of leadership. Our successes or failures, personal and collective, determine the lives of billions. Our decisions, personal and collective, determine the future of the planet.

In the spirit of responsibility, the White House and certain Congressional leaders have declared that, before beginning new offshore drilling for oil, there must be greater understanding of the environmental impact and responsibility for such endeavors. We support this approach. For, as confident as interested parties were that a disaster like this could not occur because of watertight controls and fail-safe mechanisms installed, those controls and mechanisms failed, with the horrific results we witness unfolding each day.

Until such understanding and responsibility have been determined, may God grant us all the strength to curtail the spill, the resources to support the region, and the courage to make the necessary changes so that similar tragedies may be avoided in the future.

Index